About the Author

SCOTT FLANSBURG ha[illegible] entertaining people with his astonishing abilities for more than fourteen years. People challenge him daily to calculate large math problems in his head. Scott Flansburg is "the Human Calculator," a nickname given to him by television star Regis Philbin. With his ability to add, subtract, multiply, and divide, and even to do roots in his head with amazing speed and calculator accuracy, the nickname has stuck. Scott can add the same number to itself more times in fifteen seconds than an accountant with an adding machine. For this ability, he is the *Guinness World Book* record holder for the Fastest Human Calculator.

Scott is currently working on the most important project in his life. He has founded the math company i count! and its associated event, the National Counting Bee. The vision and goal of these entities are to elevate math confidence and self-esteem for children and adults by providing educational material and awarding scholarships and prizes for math proficiency. Scott's mission is to ensure that every child and adult on the planet can add and subtract with confidence.

MATH MAGIC

MATH MAGIC

How to Master Everyday Math Problems

Revised Edition

SCOTT FLANSBURG

WITH VICTORIA HAY, Ph.D.

Perennial Currents

An Imprint of HarperCollins*Publishers*

A previous edition of this book was published in 1993 by William Morrow and Company.

HarperCollins books may be purchased for educational, business, or sales promotional use. For information please write: Special Markets Department, HarperCollins Publishers Inc., 10 East 53rd Street, New York, NY 10022.

First Perennial Currents edition published 2004.

Designed by Bernard Schleifer

Library of Congress Cataloging-in-Publication Data
Flansburg, Scott.
Math magic : how to master everyday math problems / Scott Flansburg with Victoria Hay.— Rev. ed., 1st Perennial Currents ed.
p. cm.
ISBN 0-06-072635-0 (pbk.)
1. Mathematics—Study and teaching (Elementary). I. Hay, Victoria. II. Title.

QA135.6.F73 2004
513'.12—dc22

2004040091

04 05 06 07 08 RRD 10 9 8 7 6 5 4 3 2 1

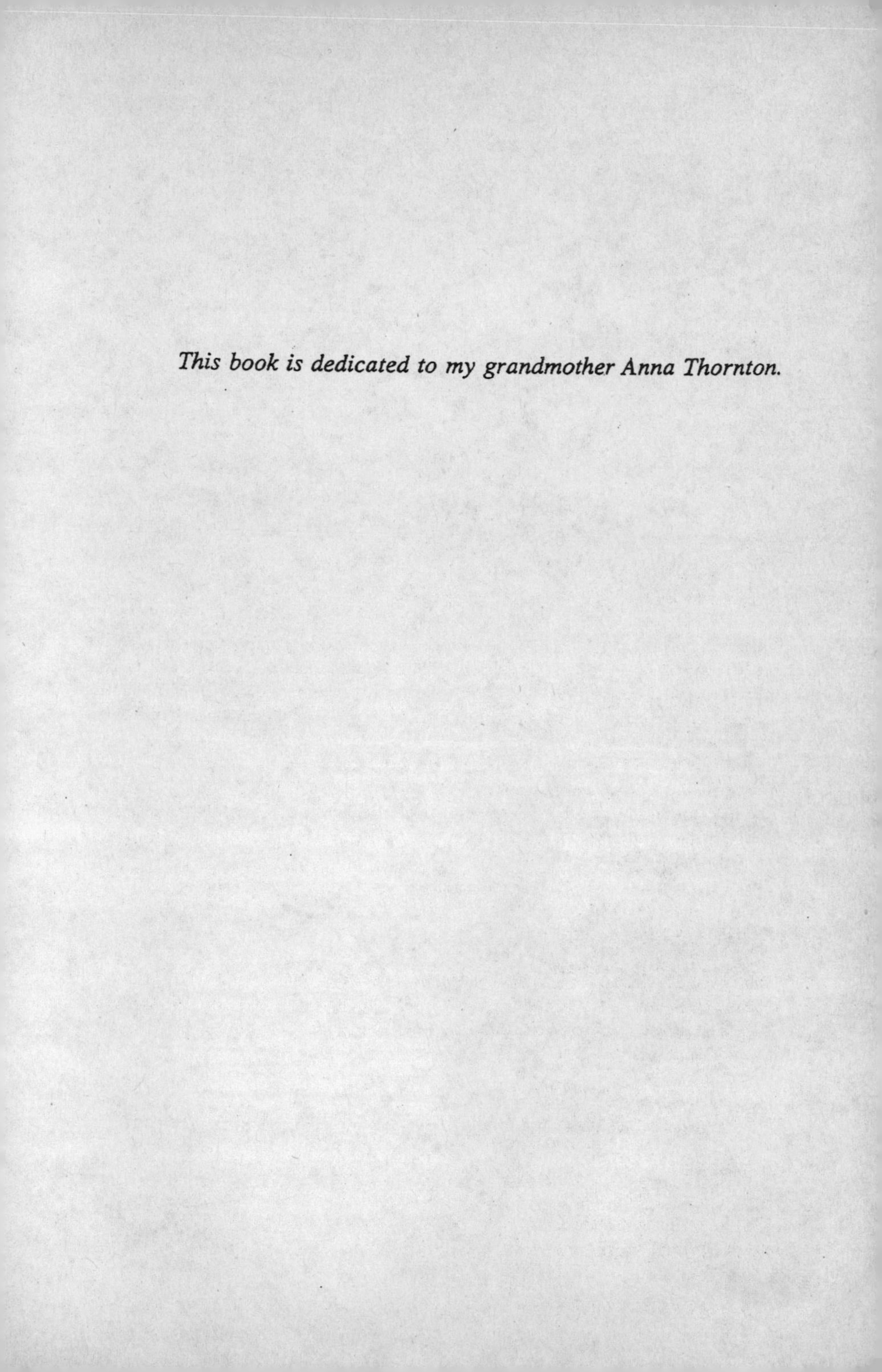

This book is dedicated to my grandmother Anna Thornton.

Don Davenport, thank you for reminding me to chase my dreams.
Jon Lovitz, thank you for helping me get started on my journey.
Kay Dean, thank you for providing me the first opportunity to share my message.
Regis Philbin, thank you for giving me my title.
Sue Colwell, thank you for believing in me from the very beginning.
Eddie Webb, thank you for putting me on the air for the first time.
Freeman Theriault, thank you for all you do for children and having 43 holes in one.
Charles Barkley, thank you for your time and generosity on my first school tour.
Mike Levy, thank you for creating my first infomercial.
Nancy Marcum, thank you for launching my first infomercial.
Mike Rowe, thank you for breaking me in at QVC.
Paul Bader, thank you for launching my first CD-ROM.
Eric and Bill, thank you for all the TV opportunities.

SPECIAL THANKS TO

Mark Abes
Rita Aguirre
Andrew Ashwood
Glen and Kim Campbell
Doug Dingman
Scott Kim
Mel and Beth Shultz
Chris Morrow
Joe Sposi
Danny and Leslie Zelisko
Dennis Haskins
Bruce Parker
Bruce Larson

Gary and Diane McCord
Mark Tarbell
Mark and Jackie Cordes
Miguel and Kelly Galeano
Vicki and Dale Jensen
Michelle and Chris Jensen
Rob and Jill Lester
Vince Marold
Anthony Santa Maria
Charles Barr
Mary LoVerde
Melanie Wicker
Stephanie Heaton
Clive and Kerry McCorkell
Maynard Small

THANKS TO ALL THE PEOPLE WHO HELP ME ON A DAILY BASIS

Andy Mckay
Jim Faughnan
Matthew Flansburg
David Dow
Kevin Hickey
Mark Thompson
Rick Buonincontri
Janet Western

I support and would like to thank Alice and Sheryl Cooper, and the staff and supporters of Solid Rock Foundation, whose goal is to help teenagers succeed through their adolescent years by promoting strong family units, good moral values, and supportive programs.

Thank you to my editor, Kate Travers.

Contents

PART III
HANDY THINGS TO KNOW

PART IV
MATH IN REAL LIFE

PART V
THE NEXT STEP

PART VI
APPENDICES

Introduction

"*Know, love,* and *believe* what you're doing—and you'll be the happiest person there is." This is the message I've been carrying across the United States and Canada for the past three years. I'm living proof that the statement is true.

In 1989, J. B. Lauchner, Sue Colwell, and I formed a company, which we called *Youth Enterprises, Inc.* Our goal was to make a difference in children's way of looking at and relating to math. We believe there are no mathematical illiterates—only people who have yet to learn to do math in a way that works for them. As with any problem in life, there's more than one way to solve a math problem.

While on our mission, we ran into a surprise. Our original program, Motivation thru Mathematics, was aimed at children. We discovered, though, that their parents and other adults were just as eager to learn our fun and creative way to manipulate numbers. Lifelong math phobias had left scars on their psyches—they had even altered their career choices to avoid math—but now they were faced with the responsibility of helping their children learn math. Cold with fear, willing to try anything, they found our program was as useful for themselves as for their children.

To our delight, too, teachers were just as excited to learn new methods they had never seen before.

Can you imagine being a teacher today? Imagine keeping children raised with remote control multichannel gratification interested in math taught with the same methods that have been used for the last one hundred years! I come into a classroom for an hour, have fun, and show some new ways of looking at math. Usually, the teachers are seeing these methods for the first time.

The responsibility for educating our children cannot be left to teachers alone. We all must take a part—not just teachers, but parents, grandparents, and corporate America. We all offer creative new methods that can reach both left- and right-brain learners, girls and boys. Everyone is capable of learning math and enjoying it.

Numbers are around us every day. There isn't a profession that doesn't use math. A person who can't understand numbers is just as handicapped as a person who can't read. And yet as a society, America approaches math in a self-defeating way.

If you take all the fallacies that we labor under—"girls can't learn math"; "people who are good with words are bad in math"; "you don't need to know math because calculators and computers will do it for you"—then add a tradition of teaching that consists of put-downs and tests guaranteed to build counterproductive tensions, you have a formula for failure. In what other class is a student asked to stand at the blackboard, in front of all the other kids, and prove whether he or she can or cannot perform! Where else do you have ten minutes to solve 20 problems, for a grade?

It's frightening, too, how little we challenge the young minds in the classrooms. Many students today can graduate from high school with only one math course: general math in the ninth grade. We demand less and less from our kids. The less we expect, the less we get.

It doesn't have to be that way. One of my most exciting experiences with children happened in Phoenix, Arizona. I had been invited to speak to a group of second-graders. I was showing them how to add left-to-right, but I was limiting my examples to two-digit numbers. They were, after all, only in the second grade, and we normally visit the third grade and up. After a few exercises, the kids said, "Let's do three-digit numbers." We did three-digit numbers. Then they were asking for four digits. Suddenly they were raising their hands faster than we could write on the board. They were screaming out the answers! Every kid in that room was excited, and not a single one was lost. They were all participating. Some of them got wrong answers, but that was all right, because they knew it was okay to be wrong.

When children are given a creative tool to overcome their own anxiety, they create enthusiasm and self-motivation. Once teachers understand that there are many different approaches to math, their enthusiasm is contagious. That's when math scores start to go up.

I greeted you with "*Know, love,* and *believe* what you're doing—and you'll be the happiest person there is." I've always enjoyed math. When I was younger, I thought all adults could do what I could do. I thought I was just ahead of my peers. I was also fortunate in having teachers in upstate New York who were very liberal. They said that if I could explain my methods to everyone, they would allow me to continue using them. I've heard a lot of horror stories in the last few years about experiences learning math—enough to help me realize how fortunate I was in receiving my math education the way I did.

I hope I can dispel the myth that math is only for nerds. After high school, I joined the Air Force "to see the world," and that's what I did. I spent three years in Japan. I learned to speak Japanese and found the culture there fascinating. But what was most important to me was that the Japanese were as intrigued by my math methods as my fellow Americans. People everywhere have a need to learn what makes math work and how to make it work for them.

While I was in the service, a friend asked me to help his son. The boy was failing in math. As I showed him new ways of doing math, I experienced great satisfaction in watching his mind and self-esteem open up. His teacher was amazed at his new abilities, and she asked me to speak to the rest of her class. That's when I began to believe in what I could do for both children's and teachers' confidence in themselves and in math. I *believe* in everyone's ability, and my own ability to touch their lives. That leads me to *loving* what I'm doing! I now have the opportunity to speak on national television, on radio, and in front of corporations. I speak to thousands of students and maybe, just maybe, I help to change their lives for the better. What more can one ask from life?

I give you this book. Remember to have an open mind. Have fun, and grow rich with confidence.

PART I

THE BASICS

Chapter Zero

CONGRATULATIONS ON TAKING the initiative in your journey toward a better understanding of math! We have all heard of the three R's: reading, writing, and arithmetic. Although only one of these disciplines actually begins with the letter *r*, it is no joke that literary and mathematical proficiency form the foundation of our education. So why do so many intelligent people have such difficulty with math? The passion and goal of my life is to end "math illiteracy." *Math Magic* is an educational tool that will give you a powerful insight into basic mathematics. Whether you are just beginning your math education or are a seasoned mathematician, this book will teach you new techniques in solving problems, thus improving your "math literacy."

Math is a language. In fact, it is the only universal language. Unlike the written and spoken word, mathematics doesn't change when you cross international borders or are trying to communicate with people from different cultures. $1 + 1 = 2$, everywhere! In English, we use our ABC's to spell words and form sentences. To do the same in Japanese we use kanji. When communicating between the two, meaning is inevitably lost in translation. Not so in math! The "alphabet" of mathematics is 0 through 9, and everyone in the world uses it. With 0 through 9 we "spell" numbers, form equations, and solve problems. In written language, if you write the exact same sentences as someone else it is called plagiarism. In mathematics, if you solve the exact same problem as someone else it is called the 100 percent correct answer. Further, it wasn't the language of any nation that put a man in space; it was the language of mathematics. For proof of this ask an old Russian physicist. . . . Although you may not understand his native tongue, he can show you the math.

Whether or not you want to guide astronauts into outer space, you will become more confident with numbers and make better decisions by using this book. I promise there won't be anything here that you can't understand and do yourself. You will gain number sense, thereby developing practical skills you can use in everyday life, from balancing your bank account to counting calories and carbohydrates, from calculating tips at restaurants to calculating your investments. You will understand math better, and you will be amazed how your new understanding will change your life.

Let us start at the beginning: zero. Zero as a concept and as a number is relatively new. Primitive counting was basic: a killed deer was a notch on a bone or stone; two notches meant two deer; etc. When counting deer, there was no zero mark on their bone or stone. Zero as a number or symbol made no sense. Back then, the ability to count was seen as mystical, magical. Zero wasn't even a thought. To this day we use a similar system of counting. A mark on the paper like this: | represents 1, || represents 2, ||| represents 3, |||| represents 4, and 5 is a slash across all four. This system is inherited from the ancients and is their way of counting. We learn this math in kindergarten. Hold up your closed fists and point a digit each time you count a number: 1, 2, 3, 4, 5, 6, 7, 8, 9, 10. This system survived from the ancients through the Egyptians, Greeks, and Romans. Each powerful civilization adopted and modified the arithmetic and geometry of the civilization before it. Math was used for measuring land. A plot of land 200 feet by 150 feet has an area of 30,000 square feet. A plot of land 450 feet by 0 feet has an area of how many square feet? The concept didn't make sense. Nor did the abstract concept of negative numbers. A duck vendor in Egypt would never ask, What is 4 ducks − 6 ducks? The concept of negative ducks makes no sense! These civilizations did not need zero as a number. Their system worked for "practical" applications like counting livestock and constructing buildings. Unfortunately, their denial of zero caused many problems. Imagine modern banking without zero. Although the concept of negative ducks makes no sense, it is very real that a bank account can become negative by overdrawing. When ideas became more abstract the ancient concepts of mathematics became inadequate to explain them.

Although the Babylonians used zero as a placeholder, like the zero in 407, we don't see zero the number until about 650 A.D. On the other side of the world, mathematicians in India adopted zero as a number. The seventh-century mathematician Brahmagupta described rules for zero. Here is a modern translation:

> *The sum of zero and a negative number is negative, the sum of a positive number and zero is positive, the sum of zero and zero is zero.*

And:

> *A negative number subtracted from zero is positive, a positive number subtracted from zero is negative, zero subtracted from a negative number is negative, zero subtracted from a positive number is positive, zero subtracted from zero is zero.*

Finally, zero was defined! And with the adoption of zero came zero's equal and opposite: infinity. Like the yin and yang, one cannot exist without the other. They are a dualistic paradox. The biggest questions in science and religion are about nothingness and eternity, zero and infinity. By embracing zero the Eastern mathematicians continued the evolution of math. Islamic and Arabic mathematicians continued this process. Al-Khwarizmi wrote *Al-Khwarizmi on the Hindu Art of Reckoning*, which uses zero as a placeholder in the positional base notation: 1, 2, 3, 4, 5, 6, 7, 8, 9, and 0. Around 1200 A.D. the famous Italian mathematician Fibonacci infused the Hindu-Arabic with the European number system, and in doing so, zero was adopted. Although he did not use zero with the sophistication of his Eastern contemporaries, he integrated the concept and thereby furthered mathematics.

Today, the concept of zero is vital in our understanding of math. Without it, we could have no negative numbers; calculus would not exist; theoretical physics would be dead; and math would be stuck in the Dark Ages. Yet despite all of the advances in our number system, we find ourselves stuck with educational tools that are outdated and fundamentally flawed. Look at the keypad of a telephone for instance. Note that zero is at the bottom and out of sequence. Same with a typewriter or computer keyboard, where zero comes after 9. The true sequence of the first ten digits is 0, 1, 2, 3, 4, 5, 6, 7, 8, 9. From this sequence all numbers are created. Zero has finally been defined in its proper place, preceding 1. It is the first digit, that from which all numbers begin. Thus, in this, the newest edition of *Math Magic*, I have added Chapter Zero. This is ground zero from which your greater understanding of math has already begun.

In addition to the exciting new techniques you will learn in this book, we would like to tell you about a national contest that will be held on November 11, 2004. With the help of sponsors and a great group of hardworking men and women, we present you with the Zeroth Annual Counting Bee. That is right: not the first annual, because it takes more than one to make it an annual! From each of the fifty states we will be flying out two individuals and both their families to join us in the National Counting Bee. The National Counting Bee is an event built upon the dream that we can all be "mathletes," and put an end to math illiteracy. The first category is for nine-year-olds. The second category is open to anyone who wishes to enter. The rules are simple and it is very easy to enter. We are awarding tens of thousands of dollars' worth of prizes and scholarships, so log on to www.1800additup.com or call 1-800-ADD-IT-UP and find out more of the details.

So, why do so many intelligent people have such a difficulty with math? The answer is you don't have to! Your choice to read this book is the first step in becoming a math whiz. Congratulations—you have finished Chapter Zero.

1 Five Keys to Human Calculating

MATH BECOMES A LOT easier when you understand a few basic principles, some or all of which you probably learned in school. But because most of us are taught to add and subtract backward—right to left, instead of left to right—the real significance of these principles and their use may have escaped you. Let's start with a simple question:

WHAT DO NUMBERS REALLY MEAN?

A number's meaning is found from the order in which we write down its digits. The *digits* of a number are those figures that, like letters in a word, make up the total.

For example, 8,976 contains four digits.
↑↑↑↑
DIGITS

The value of a number depends on how many digits it contains. A single digit means its value is somewhere between zero and nine.

A one-digit figure is called the *ones* or *units* when it's part of a larger number, and we write it on the far right side of the number: 8,97**6.** When we write the number there, we say we are placing it in the "ones column" or "units column." In this number, the *6* means there are six ones in 8,976.

A two-digit figure stands for ten or more. It means you can count as many *tens* in the number as the left digit says. For example, 20 holds two tens; 28 has two tens and eight ones. In a larger number, we write the digit that counts the tens just left of the *ones* digit: 342 has four tens and two ones. Using our example number, 8,9**7**6, picture what this number contains by counting to the *tens* digit (the *7* in 8,976) by tens: ten, twenty, thirty, forty, fifty, sixty, seventy. It has seven tens and six ones: 8,976.

Three digits stand for one hundred or more. After seeing how tens work, you get the idea: The number 136 contains one *hundred,* three *tens* (or "thirty"), and six *ones* (the digit "six"). And 554 stands for five hundreds plus five tens ("fifty") plus four ones ("four"). We write hundreds in the third column from the right, just to the left of the tens. Using our example number again, 8,976, we find it contains nine hundreds: 8,**9**76. As we work from right to left, we separate each three digits with a comma: 1,876,325. This helps make the larger numbers easier to read.

Thousands are written in the fourth column from the right (to the left of the *hundreds*), and of course they stand for how many thousands a number contains. The enormous 1,876,325 has six thousands, found in the fourth from the right column. In our example, the number of thousands is eight: **8**,976.

So it goes toward infinity. As you read from right to left, you come to ten thousands, hundred thousands, millions, and so forth. The figures in each of a number's columns are called its *place values:* 8, 9, 7, and 6 are place values. Remember that term.

THINKING FROM MOST TO LEAST IMPORTANT

When students first learn to add, they start with ones. Maybe that's why in school we learn to add larger numbers right to left—from ones to tens to hundreds—instead of the more logical way, from left to right—from the biggest digits to the smallest ones. But the way you learned to calculate in the second grade is not the only way, or even the best way, to add and subtract.

Instead, you may find it easier to work from left to right, in the same direction that you read plain English. If you understand that the first digit you see—the figure on the left—means that a number contains so many thousands, or so many hundreds, or so many tens, you can estimate at a glance how many thousands (or hundreds, or tens, or whatever) the sum of several numbers will contain.

Think of it in terms of dollars and cents. Suppose you wanted to add

Big numbers	*Dollars and cents*
4,823	$48.23
2,762	27.62
1,827	18.27

Pennies and dimes don't make much difference when you're dealing with twenty- and fifty-dollar bills. In larger numbers, the ones and the tens are like pennies and dimes.

So, start with the important numbers! Look at the digits on the left: 4 plus 2 plus 1 equal 7. In the first list of numbers, that stands for thousands: 7,000. You know your answer can't be less than 7,000, and—sneaking a peek at the hefty numbers in the hundreds column—it probably won't be more than 9,000. Over in the money list, the 7 stands for seven tens or 70. The big bucks there add up to no less than $70, and a few pennies on the right of the decimal point are just pocket change.

The Human Calculator's System for Marking Place Values

In this book, I will use one or more 0s in parentheses to mark places that I want you to keep in mind. For example, if we're working with digits in the hundreds column and we don't want to forget the tens column and the ones column, I will write the number as 1(00). If we're working in the thousands column, I might write something like 2(000). The 0s in parentheses mark the places of the other columns.

Pretty easy to understand, isn't it! You've just discovered two of the five keys to becoming a human calculator, and it hardly hurt at all. The idea that the value of a number falls into different columns of digits is called *place value*. Using that idea to guess at how much a bunch of numbers adds up to is called *estimation*.

USE YOUR MEMORY

Your brain and your desire are the only equipment you need to become a human calculator. Most people use about 10 percent of their brain's capability. Think of what would happen if you used *only* 10 percent of any other organ in your body. If your legs worked at just one tenth of their potential, you would have to use a wheelchair or a cane, and that would be considered a great tragedy.

You can train your mind to work at a higher capacity by *using it more*. That means reading and writing and searching for answers to the questions your natural curiosity brings to you.

Where math is concerned, remember that your brain can store things like a computer. With practice, you can train yourself to hold numbers in your memory—upgrading your mental powers by two. How? By using your imagination to visualize numbers as you add, subtract, and multiply them in your mind. Form a clear picture of the problem in your mind, and work out the solution in your head.

Sound hard? It's not, really. In fact, it's nearly impossible *not* to visualize numbers, even if you try not to. As you add or subtract from left to right, you will work from a *base number* that gives you a rough estimate of the answer. Commit this easy figure to memory—it's usually shorter than your phone number—and you will find it surprisingly easy to figure the exact total in your head later.

PRACTICE, PRACTICE, PRACTICE

Sure, your piano teacher told you the same thing. But that old saying "Practice makes perfect" has a lot of truth. Three factors will improve your math ability:

1. Simple strategies to add, subtract, multiply, and divide
2. Memory
3. Practice

Even though the strategies are simple to learn and use, you can't expect to read about them once and then pull them out of your memory at will. Instead, you need to make them as natural as breathing. You do that by practicing—every chance you get.

Practice doesn't necessarily mean you have to spend hours sweating over a workbook. Math, as we have noticed, is everywhere. Use your imagination to see the numbers around you, and play with them as you go. On a trip to the grocery store, add up the prices of the goodies you toss in your shopping cart. When you arrive at the checkout line, subtract all your discount coupons from your total and add the tax. With some practice, you'll soon have the tab before the cashier can add it up on the cash register! In your head, keep track of the family car's mileage by watching the odometer and the gallons of gas the car guzzles. Down at the beach, estimate how many surfer-size waves roll into shore during a good day by counting the number of good breakers in 10 minutes and multiplying by 6 and then by 24.

BE CREATIVE

Math is more than rote memorization and practice. Math is a creative activity, like drawing or writing or playing basketball. In fact, one of the best definitions of creativity that I have heard says that "creativity is play."

Yes! The things in life that are the most fun are creative things. I am always looking for new ways to manipulate numbers, for new relationships between figures, for quirky or interesting things to do with math. All the time, numbers are going through my head. I'll take a goal number, such as 1,000, and try to figure out what numbers I could multiply and then add together to come up with that goal. Finding new strategies and shortcuts is like a game, as challenging and exciting as painting a picture or working out new plays for a football team.

A touch of creativity will help you see the fun in practicing math as you go through each day. And it will also show you that numbers are really nothing more than common sense. Math is far from dull and difficult. It's fun, exciting, and sometimes even kind of crazy.

THE KEYS TO HUMAN CALCULATING

In this chapter, you have discovered the five keys to mental math.

1. Understanding what numbers mean
2. Thinking about numbers forward instead of backward
3. Memory

4. Practice
5. Creativity

Now let's take those keys and use them on the front door to mathematics: *addition.* Adding is the most basic part of arithmetic. Learn it well, and it will open the house of mathematics to you.

2 Addition

BASIC STEPS OF LEFT-TO-RIGHT ADDITION

THERE'S ONLY ONE POINT to doing arithmetic: to get the right answer. It doesn't matter *how* you get the right answer, as long as your method works. It works if you get the right answer every time you do it.

If you've heard me on television or radio, you know that I can add 123 plus 226 plus 121 plus 214 as quickly as the participants from the audience can total those numbers on their calculators—the answer is 684. I haven't memorized the problem and the answer, and it is not a trick. I've learned to solve math problems fast by combining estimation with left-to-right addition and holding a few figures in my head as I go.

You can learn to do this, too. Let's try a strategy on those numbers:

$$\begin{array}{r} 123 \\ 226 \\ 121 \\ \underline{214} \end{array}$$

Notice that the digits in the hundreds column are in italic type; the digits in the tens column are boldface; and the digits in the ones column are in ordinary type.

Add from Left to Right

Remember what we learned in the last chapter? The most important numbers in a problem are the digits on the far left, because they're the largest. The largest numbers have the biggest impact on your answer. So start with the important digits, the hundreds (the ones that are italicized).

Add the 2 and the 1. Because they're hundreds, you're really adding 200 plus 100, which gives you **300.** Now you've started a *base number*—the figure you'll work with in your mind as you go. We'll mark the place values that we have yet to work out with 0s in parentheses: 3(00).

Add the rest of the figures in the hundreds column to the base number of 3(00): 2 (which is really 200) plus 3(00) makes 5(00); 1(00) plus 5(00) is **6**(00).

Now, we're going to keep building on our base number of 6(00). Move over to the tens column. Start with 2 (which is really 20, or, in our system of holding place values open, 2[0]) and add it to the base number of 6(00), giving you 62(0). The next number down is also 2(0); add it to 620 and you get 64(0). Add the next number down, another 2(0) for 66(0). And finally add on the 1(0) to get a new *base number,* **67(0).**

Now hop over to the ones column next. The first number is 3, which we know stands for plain old one-digit 3. Add it to 670 and change the *base number* to 673. Next is 6, which added to 673 equals 679. The next number is 1, raising the base number to 68(0). When we add the last number, 4, we get the new base number—and **the answer: 684.**

The answer to an addition problem, by the way, is called the *sum*.

As you were following this simple procedure, you may have picked up right away on a handy fact. The final answer *could not be less than 600!* It would have taken awhile to figure that out if you worked the problem the traditional way, from right to left.

Now, you understand that when a column of numbers adds up to more than 9, the amount more than 9 flips over to the left-hand column. This number can be used to tell that an answer *will not be more than* a certain amount.

Let's see how that works with the following problem.

242
163
431

Start with the hundreds column: 2(00) plus 1(00) is 3(00), plus 4(00) gives us a base number of **7(00).** Okay. We know the answer cannot be less than 700.

"Carrying" from Column to Column

In the tens column, the base number is 7(00) plus 4(0) equals 74(0). Next we add the 6(0) to get . . . uh oh. The sum is 10(0) when you add 6(0) to 4(0). You can't write the two-digit number "10" in the one-digit tens column. What to do?

Well, 60 plus 40 is 10 tens. What's happened? You've collected enough 10s to make a hundred—that is, 10(0), or 1(00), or 100. So, you add a hundred to the hundreds column in your base number: 700 plus 1(00) gives you 800. In this case, no tens are left after adding the 6(0) and 4(0)—the number in the tens column is now 0. So you add nothing to the tens column in the *base number*. The new base number is **8(00).**

Now, the tens left in the problem after adding the new 1(00) to the hundreds column are simply added to (00) in the tens column. Pretend, for example, that instead of 40 we were adding *50* to 60. Our base number from the hundreds was 7(00); then 50 plus 60 would make 110, or 1(00) and 1(0). We would have added our 1(00) to our old base number to get 800. We then would have picked up the leftover 1(0) in the tens column, to get 810. Remember: When you flip over to the next column, *clear the old column to zero first, before you add leftover digits.*

Back to the original problem: One more number remains in the tens column: 3(0)—8(00) plus 3(0) equals **83(0).**

Finish with the Units Column

All right! We're ready for the ones, and we know that, no matter what, the answer will not be more than about 860 because three digits in the ones column must be less than 30. Add on the little numbers on the far right: 830 plus 2 is 832, plus 3 is 835, plus 1 is **836.** The last base number **(836)** is the answer.

Now You Can Do It

Add from left to right, instead of the old-fashioned way, and see how much faster you can find the sums of these numbers:

23	92	146	136	358
46	24	273	992	875
		121		

1,243	2,576	29	8,703	9,272
2,345	5,436	33	2,961	5,498
1,532		28	343	27
		92		2,309
				8,565

Estimating the Answer

Going back to our example on page 21, whose answer was 836, how did I know that the solution to this example would not be more than 860 once I'd finished with the tens and gotten a base number of 830? Easy.

I know the largest numbers that can possibly appear in the ones column are 9s. There are three numbers in the problem, so even if each of them had ones of 9, the highest possible total of ones digits is 27. Three 9s are 27. Add 27 to 830, and the answer is 857. Try it:

$$\begin{array}{r} 249 \\ 169 \\ \underline{439} \\ 857 \end{array}$$

No question of it! Our problem can't add up to more than 857. Chances are you won't ever have three 9s, but you know the highest possible total is 857, or *about* 860.

Now you know how to make your teacher and your friends think you're brilliant. Halfway through a problem, strike a brainy pose and murmur something like "Hmm. The answer lies between 800 and 860." You *will* sound impressive!

This technique is called estimation, and it comes in handy in a lot of other situations. Often there's no need to calculate an answer right down to the penny. Let's say that you want to figure how much to tip the efficient waiter who served you and your three friends. The lunches cost $4.28, $7.46, $5.53, and $6.21. Adding from left to right, you see the total will not be *less than* $22.00. A glance at the numbers in the tens column suggests the bill won't run much over $23.00. You don't need to know that the total is $23.48 to realize you need to tip a little more than 20 percent of $20.

Now You Can Do It

A glance at the hundreds and tens digits of these numbers tells you their sum is larger than what but smaller than what? 289 plus 444 plus 562 plus 867.

MAKING IT WORK ON BIGGER NUMBERS

So far, this has been pretty easy. But what if the numbers are bigger? What if there are more of them? Will this strategy keep working? Let's try it out on a few more serious challenges.

First, some larger figures. Let's add:

8,461
7,353
6,127

Start from the left, in the thousands column: 8(000) plus 7(000) is 15(000), plus 6(000) makes a base number of **21**(000). Now move to the hundreds: 21(000) plus 4(00) is 21,400, plus 3(00) is **21,7**(00). Add 1(00) to arrive at **21,8**(00). Now add the tens: 21,8(00) plus 6(0) equals 21,860. Six (tens) plus 5 (tens) will give us more than 10 (tens)—get ready to carry some numbers. Think: 6(0) plus 5(0) is 11(0). That's 1 hundred plus 1 ten, or 1(00) plus 1(0). Add another 1(00) to the hundreds column: 8(00) plus 1(00) equals 9(00). One ten, or 1(0), is left in the tens column, making the new base number **21,91**(0). We still have two more tens to add, for **21,93**(0). Now all we have left to count up are the ones: 21,91(0) plus 1 is 21,911, plus 3 is 21,914. Add 7 and you get 11 ones, or 1 ten—1(0) and 1 unit. Carry that 10 over to the tens column and clear the ones column to zero: 3(0) plus 1(0) is 4(0), changing the base number to **21,94**(0). Now add the leftover unit and you have the answer: **21,941.**

Picturing the Strategy in Your Mind

Four-digit numbers are about the largest figures most people can add in their heads, although the principle works on any number you can conjure up. Describing this procedure in writing makes it look more complicated than it is. Actually, it helps to think about figures this size by forming pictures of them in your mind. So, picture what we just did like this:

```
     8(000)
 +   7(000)
 +   6(000)
 ----------
    21(000)
 +    4(00)
 +    3(00)
 +    1(00)
 ----------
   21,8(00)
 +     6(0)
 +     5(0)
 ----------
   21,8(00) + 10(0)  =  21,9(00)
                      +     1(0)
                      ----------
                        21,91(0)
                      +     2(0)
                      ----------
                        21,93(0)
                      +        1
                      ----------
                          21,931
                      +        3
                      ----------
                          21,934
                      +        7
                      ----------
                        21,93(0) + 1(0) + 1
                   =  21,941
```

Working the Strategy on a Long Column

It's just as easy to add up a long string of numbers. Try your new left-to-right skill, for example, on this one:

22
18
45
62
94
39
55
71
86

Remember to start on the left—in this case, in the tens column. From the top: 2(0) plus 1(0) is **3**(0), plus 4(0) is **7**(0). Adding 6(0) will push our tens column into the hundreds, because 7(0) plus 6(0) is 13(0). We'll do it again when we add 9: 3(0) plus 9(0) makes 1(00) and 2(0). So, add 1(00) to the 1(00) in the hundreds column of the base number, 130, to get 2(00), and then add the leftover 2(0) to 2(00) for a new base number, **22**(0). Add 3(0) to get **25**(0). Pick up another hundred when you add 5(0): 5(0) plus 5(0) is the same as 10(0), which is the same as 1(00). Just as you did before, add the new 1(00) to the old hundreds column, 2(00), to get **3**(00). Add 7(0) for **370**. Another 8(0) gives us one more hundred, like this: 7(0) plus 8(0) equals 15(0), or 1(00) and 5(0). Add the 1(00) to the hundreds column and get 400, and keep the 5(0) for a base number of **450.**

Now jump over to the ones column and start adding units to the base number: 45(0) plus 2 is **452,** plus 8 is 45(0) plus 1(0) or **460.** Add 5 to get **465,** plus 2 is **467.** You get a new ten when you add 4: 467 plus 4 equals 4(60) plus 1(0) plus 1, or **471.** Adding nine gives us another new ten and with no units left over: 1 + 9 makes 1(0), added to 471 is **480.** And 5 to get **485,** plus 1 equals **486.** We gain one more ten when we add 6: 6 plus 6 makes 12, the same as 1(0) plus 2; 48(0) plus 1(0) is 49(0), and the leftover 2 brings us to the answer: **492.**

Picturing the Strategy for a Long Column

How does this procedure look to you when you try to visualize the numbers in your mind? Here's how it looks to me:

2(0)
\+ 1(0)
\+ 4(0)
7(0)
\+ 6(0)
1(00) + 3(0) = **13**(0)
\+ 9(0)
2(00) + 2(0) = 22(0)
\+ 3(0)
25(0)
\+ 5(0)
30(0) or 3(00)
\+ 7(0)
37(0)
\+ 8(0)
3(00) + 1(00) + 50
=
45(0)

Add the units:

```
 450
+  2
 452
+  8
45(0) + 1(0) + 460
             +   2
             +   5
               467
             +   4
             46(0) + 1(0) + 1 = 471
                              +   9
                          47(0) + 1(0) = 480
                                       +   5
                                       +   1
                                         486
                                       +   6
                                         48(0) + 1(0) + 2 =
                                         492
```

APPLYING THE KEYS TO HUMAN CALCULATING

As you can see, the trick to being fast and accurate is really just a matter of common sense and of using your five keys to mental math.

Understand what numbers mean. Keep thousands, hundreds, tens, and units straight, and remember that whenever a column has more than 9, it will flip the column to the left up in value.

Think about numbers forward instead of backward. Figuring from left to right makes it easy to estimate an answer. It's also a logical way of looking at math.

Develop your memory. Getting a correct answer depends on your knowing the math facts. Yes—all those addition, subtraction, multiplication, and division tables. You need to know the math facts through the tens, and preferably through the twelves. This is not as hard to learn as it looks, if you train yourself to exercise your memory. Whenever you get a chance—at the grocery store, down at the fast-food restaurant, in the gas station—add, subtract, multiply, and divide the one-digit numbers you see. If you can't remember the answer quickly, add it up on your fingers. Pretty soon you *will* remember.

Practice. Give your brain a mental workout each day, just as a healthy person works out his or her body. Your brain is like your muscles. The more you use them, the faster they get and the better they work.

Be Creative. Don't hesitate to try new strategies and shortcuts. If they work, find out how and why, and look for ways to use them to make math easier and faster for you.

SHORTCUTS AND HELPFUL HINTS

When you free up your mind for a few creative approaches, you find all sorts of tricks and gimmicks to speed up your mental calculating.

Take, for example, the fact that you can add or multiply numbers in any order. That is, 2 + 5 + 8 is the same as 5 + 2 + 8, which is the same as 8 + 2 + 5. Since multiplying is just another form of adding, 2 × 5 × 8 equals the same thing as 5 × 2 × 8 or 8 × 2 × 5. You can put this reality to good use. Because some numbers are naturally easier to add or multiply than others, you can use them out of order for quicker results.

Ten and its multiples, for example—10, 20, 30, and so on—are especially easy. When you're adding a long series of numbers, you don't have to take them from the top and go straight to the bottom. Instead, look for sets of numbers that add up to 10, and keep track of the 10s as you go.

Look back at the long problem under "Working the Strategy on a Long Column." Notice that the left column has four sets that make 10: 4 + 6, 1 + 9, 3 + 7, 2 + 8. Four 10s make 40. Since we started in the tens column, that means 40(0) or 4(00). The only number left is 5(0). Add it to 400 and you get an instant base number: **450.** Now add up the units the same way: 2 + 8, 5 + 5, 4 + 6, 9 + 1. Four 10s again: 4(0). In the units column, we don't have to add any 0s. The digit 2 is left, so we have 42 units, or four 10s—4(0)—and 2 units. Add 4(0) to the base number 45(0) to get **490** and then add the 2 for the answer: **492.**

Now *that* was fast! Creativity plus understanding equals speed.

Other sets of figures are easy to spot. You can, for instance, add by multiplying. Actually, multiplying is just a shortcut through addition. Three times 9, for example, means the same as 9 + 9 + 9. It also means the same as 3 + 3 + 3 + 3 + 3 + 3 + 3 + 3 + 3. If you notice three nines in a column, think "3 × 9 = 27." Several 3s should make you think of, say, "4 × 3 = 12." It's faster by far than thinking "3 + 3 = 6 + 3 = 9 + 3 = 12."

If you don't need a specific figure but only want to know approximately how much something adds up to, try rounding off numbers in a problem. For example:

$$\begin{array}{r} 98 \\ 11 \\ \underline{52} \\ 161 \end{array} \quad \text{is approximately} \quad \begin{array}{r} 100 \\ 10 \\ \underline{50} \\ 160 \end{array}$$

Even though Samuel Johnson once said, "Round numbers are always false," there are times when a round number is "good enough for government work." And they do offer an easy way to check your math.

How do you round off numbers? If the digit on the right is 5 or more, add 1 to the column you're working with. If it's 4 or less, let the column stay as it is.

54	rounds off to	50	
56	to	60	
472	to	470	
591	to	600	(rounded to the nearest hundred)
350	to	400	
4,631	to	4,600	(rounded to the nearest hundred)
4,631	to	5,000	(rounded to the nearest thousand)
23,862	to	23,860	(rounded to the nearest ten)
23,860	to	23,900	
23,900	to	24,000	

Let's add up these figures and see how big the difference is. The first column comes to 82,407. The second column totals 82,940. The amount that has been estimated is only 533 off the precise figure, not much when you're talking tens of thousands.

Speaking of small change, when you're rounding off dollars and cents, count 50 cents or more as a dollar and drop 49 cents or less.

$27.78	is about	$28.00
$ 5.20		$ 5.00

Here's one last creative trick to smooth your way to high-speed mental math. You can solve a difficult-looking addition problem in a snap by simplifying the numbers in your head. You do this by adding or subtracting small numbers to the figures you have to total. Try it:

$$\begin{array}{r} 97 \\ +\ 84 \\ \hline \end{array}$$

Add 3 to 97 to get 100, which is easy to work with. Now, 100 + 84 = 184. Subtract the 3 you added: 184 − 3 = **181.**

Or, if you prefer, add 3 to 97 to make 100. Then subtract the 3 from 84, giving you 81. Add 81 and 100 for the answer, again *181.*

This works because the answer to an addition problem will remain the same whenever you add *and subtract* the same amount. Thus, you can make a horrible-looking example very simple:

$$\begin{array}{r} 14{,}763 \\ +\ \ \ 999 \\ \hline \end{array}$$

The easy way: 999 + 1 = 1,000

$$\begin{array}{r} 14{,}763 \\ +\ 1{,}000 \\ \hline 15{,}763 \end{array} \qquad \begin{array}{r} 15{,}763 \\ -\ \ \ \ \ \ 1 \\ \hline \mathbf{15{,}762} \end{array}$$

Keep your eyes and ears tuned for easier ways to figure. With practice, you'll start to see them everywhere.

Now You Can Do It

Use all the shortcuts you know to add these numbers in your head:

14	28	288	987	102
16	32	926	452	876
82	57	342	374	743
98	74	613	451	132
15	21	353	576	224
	82			525
	79			367
				222
				567

3371	$14.01	$9.99
343	1.29	.24
2414	2.49	.37
8743		.89
9876		

3 Cross-Multiplication

THERE ARE SEVERAL WAYS to break down a multiplication problem to make it amazingly easy to do. These strategies are fun, and because most people think multiplication is so hard, you can amaze your friends, teachers, and coworkers after mastering a few tricks.

The explanations only look complicated, but once you see the beauty of these techniques, you'll realize that they are really very easy to work. In fact, you can multiply most three-digit numbers in your head.

Remember that multiplying is really a form of adding, something we do naturally. Most people are adding all the time in their minds. When we say 9×3, we're really just taking a shortcut to arrive at the same answer as $9 + 9 + 9$. People who don't know the multiplication table very well will think, "$9 + 9 = 18$, and 9 is . . . umh [they count on their fingers, also a form of adding] 19, 20, 21, 22, 23, 24, 25, 26 . . . sure! 27." Obviously, this is the hard way to add—so, you can see it's worth learning your "times" through 12s.

The number you are multiplying, by the way, is called the *multiplicand.* The number you want to multiply it by is called the *multiplier,* and the answer to a multiplication problem is the *product.*

$$\begin{array}{rl} 2 & \leftarrow \text{MULTIPLICAND} \\ \times\ 4 & \leftarrow \text{MULTIPLIER} \\ \hline 8 & \leftarrow \text{PRODUCT} \end{array}$$

RIGHT-TO-LEFT CROSS-MULTIPLICATION

Our first multiplying strategy is called *right-to-left cross-multiplication.* Let's try an example slowly.

$$\begin{array}{r} 14 \\ \times\ 12 \\ \hline \end{array}$$

To get the last digit in the answer, multiply the two numbers on the right. Multiply 2 × 4. Write the answer, 8, as the rightmost (ones) digit in the problem's answer.

$$\begin{array}{r} 14 \\ \times\ 12 \\ \hline 8 \end{array} \qquad \textbf{THINK: } 2 \times 4 = 8$$

Now, to get the middle digit we'll cross-multiply both sets of numbers, and then add them together. Multiply 1 × 2 (the answer is 2) and 4 × 1 (the answer is 4). Add the answers to get the middle digit (tens) in the problem's answer.

$$\begin{array}{cc} 1 & 4 \\ & \times \\ 1 & 2 \end{array} \qquad \begin{array}{r} 14 \\ \times\ 12 \\ \hline 68 \end{array} \qquad \begin{array}{l} \textbf{THINK: } 1 \times 2 = 2 \\ \qquad\quad 4 \times 1 = 4 \\ \qquad\quad 2 + 4 = 6 \end{array}$$

To get the last digit, which in this problem will be hundreds, just multiply the left digits: 1 × 1 = **1**. Write it down in the hundreds column.

```
  14    THINK: 1 × 1 = 1
× 12
 168
```

"Wow, Scott! That was easy." I hear those words every time I explain this method. You will be delighted at how much easier it gets with practice.

Try it on a couple of three-digit numbers.

```
  621
× 584
```

All right. It looks like a nightmare, but don't be afraid of it. Start on the right side: 1 × 4 = 4. Couldn't be simpler.

```
  621    THINK: 1 × 4 = 4
× 584
    4
```

Now, multiply crisscross: 2 × 4 = 8; 1 × 8 = 8. Add the answers to get 16.

```
6  2 ↖ ↗ 1       621    THINK: 2 × 4 =  8
      ╳
5  8 ↙ ↘ 4     × 584           1 × 8 =  8
                   4                   16
```

Now, you'll remember what we said about *place value* in Chapter 1. Going from right to left, the digits in a number represent *units, tens, hundreds, thousands,* and so on. We already have the rightmost digit in our answer—the 4—and it occupies the units column. The next digit in the answer will occupy the tens column. But 16 has two digits. We can put the 6 in the tens column, and that will give us the second digit in the answer. But what about the 1?

That 1 in the 16 is actually a *hundred.* If we were just to write the 16 next to the 4, we'd get this number: 164. To remind us that the 1 is really a hundred, we could write it in the system we developed in Chapter 1: 1(00).

Meanwhile, we're not finished with the problem—we still have some other numbers to multiply, and they will give us the digit that goes in the hundreds column. When we finish multiplying for the next step, we will have the 1(00) left over from multiplying 10s. We will simply *add* that 1(00) to the result of the next step. This procedure is called *carrying.* Watch this:

```
  621     THINK: WRITE 6, CARRY 1(00)
× 584                  (1) 621
-----                    × 584
   64                    -----
                            64
```

So far, we've multiplied with the right and the middle digits on our top number. Now move over to the top left digit and start crisscross multiplying and adding. Go $6 \times 4 = 24$; $5 \times 1 = 5$; $2 \times 8 = 16$. Add those answers, $24 + 5 + 16$, for a total of 45. Now add the 1 you carried: $45 + 1 =$ **46**. Once again, we'll have to write down one digit and carry the other.

```
6 2 1          621    THINK: 6 × 4 = 24
5 8 4        × 584           5 × 1 =  5
                64           2 × 8 = 16
                      24 + 5 + 16 = 45(00)
This is the carried 1(00):        + 1(00)
                                   46(00)
```

Write down the hundreds digit:

```
  621    THINK: WRITE 6, CARRY 4(000)
× 584           (4) 621
  664             × 584
                    664
```

Now, understand that because we were multiplying hundreds, the 6 in 46(00) is really 6(00) and the 4 is really 4(000). We're carrying the 4 into the thousands place.

One more crisscross: 6 × 8 plus 2 × 5. That's 48 plus 10, or 58. Add the 4 that you carried to get **62**, and once again write down the right digit—the **2**—and carry the left digit—the **6**.

```
6 2 1          621         THINK:  6 ×  8 = 48
5 8 4        × 584                 2 ×  5 = 10
               664               48 + 10 = 58
                                          + 4
                                       62(000)
```

Write it down:

```
  621     THINK: WRITE 2, CARRY 6(0,000)
× 584                          (6) 621
 2664                              584
                                  2664
```

We carry the 6 into the ten-thousands place. Now, the last step! It repeats the first step, only on the far left instead of the far right. Multiply 6 × 5, to get 30. Remember that you carried 6—just add it to the 30, which gives you **36**. Write it down, and you have the answer:

```
6  2  1          621        THINK: 6 × 5 = 30
|
5  8  4        × 584                      + 6
              362664                  36(000)
```

The answer is **362,664!**

"Gee, Scott. You thought that was easy? How am I supposed to do that in my head?" Well, it *is* pretty simple, as long as you have a pencil and can write down the answer as it builds from right to left. You will need some practice before you can do that as mental math, though.

But here's another strategy that you probably *can* do in your head—again, with a little practice. It works like addition, from left to right.

LEFT-TO-RIGHT CROSS-MULTIPLICATION

It's easiest to see how this works on a couple of two-digit numbers. Let's try:

$$\begin{array}{r} 36 \\ \times\ 24 \\ \hline \end{array} \qquad \begin{array}{cc} 3 & 6 \\ \updownarrow & \\ 2 & 4 \\ \hline \end{array}$$

We start on the left—in this case, with the tens column—to build a **base number.** We are going to multiply 2 × 3, which will give us **6**, but we must remember that we are *really* multiplying 3(0) × 2(0). That means our base number of **6** takes with it those two magic zeros, and so it is *really* **6**(00).

Now multiply 2(0)—the tens digit in the bottom number—times 6, the units digit in the top number.

$$\begin{array}{cc} 3 & 6 \\ 2 & 4 \end{array} \qquad 2(0) \times 6 = 120$$

Add that to our base number, **6**(00), to get a new base number, **72**(0).

All right, now move over to the units digit in the bottom number, which is a 4. Multiply the tens digit in the top number, 3(0) times 4:

$$\begin{array}{cc} 3 & 6 \\ 2 & 4 \end{array} \qquad 3(0) \times 4 = 12(0)$$

Add it to the base number of 720 for a new base number, **84**(0).

And finally, multiply both units digits, 4 times 6:

$$\begin{array}{cc} 3 & 6 \\ 2 & 4 \end{array} \qquad 4 \times 6 = 24$$

Add the result to the base number of **84**(0), and this should give us our **answer**:

$$\begin{array}{r} 84(0) \\ +\ 2\,4 \\ \hline 86\,4\text{, or } \mathbf{864} \end{array}$$

Do you see the principle? Multiply the top number by the digits, going left to right, in the bottom number. Add up the answers in your mind.

Here's a faster way to do left-to-right multiplication with two-digit numbers. Let's take 36 × 24 again. Remember that in our basic example above we took each digit one at a time. If you prefer, you can multiply 36 × 2(0) and then multiply 36 × 4. This way, you add the whole top number by each of the digits on the bottom. Add the results and you'll have the answer:

36 × 2(0) = 720
36 × 4 = 144
Add the two: **864**

Let's try it on the three-digit example we used above:

621
× 584

Once again, we start on the left to build a **base number,** the same way we did with adding. Remember that when you start with that 6 and that 5, you're in the *hundreds* column: 600 and 500, or 6(00) and 5(00). When you move over to the 2 and the 8, you're in the *tens* column, so those numbers are really 20 and 80, or 2(0) and 8(0). And the digits on the far right, 1 and 4, are simply ones, 1 and 4.

Here goes: 6(00) × 5(00) is **30**(0,000), our new base number. Why? Because when you multiply a hundred (two zeros) by a hundred (two zeros), you get ten thousand. We show that we've passed the ten thousands mark in this example by writing four zeros.

621 **THINK:** 6(00) × 5(00) = **30**(0,000)
× 584
300,000 (Keep it in your head.)

Now, visualize the first two digits in the multiplier and the multiplicand and cross-multiply: 6(00) × 8(0), for 48,(000). Add that to the base number, 300,000, for a new base number, **348,**(000). And again: 2(0) × 5(00) = 10(000). Add that in to get the new base number, **358,**(000).

```
   621   THINK: 6(00) × 8(0) = 48,(000)
 × 584          2(0) × 5(00) = 10,(000)
                               58,(000)

   58(000)   Add it to 30(0,000)
+ 30(0,000)
  358,(000)
```

The next step is to multiply 6(00) × 4, for 2,4(00). What are we doing here? We're picking up the ones unit in the multiplier (58*4*). Add the result to the base number, making it **360,4**(00).

Now go to the ones unit in the multiplicand and multiply it times the hundreds unit in the multiplier. In other words, take 1 × 5(00), which is 5(00), and add that in, for a new base number of **360,9**(00). Here's what we did in these two steps:

```
6  2  1          360,4(00)
 ╲ ╳ ╱               5(00)
5  8  4          360,9(00)
```

Now visualize the last two digits in each number and cross-multiply them: 2 × 4 and 1 × 8. Figure 2(0) × 4 equals 8(0), changing the base number to **360,98**(0). Then multiply 1 × 8(0) to get another 80, add it to the base number, and get **361,060**.

```
6 2 1
   X
5 8 4
```

```
  621    THINK: 6(00) × 4 = 2,4(00)
× 584             + 358,(000)
                   360,4(00)
       + 1 × 5(00) = 5(00)
                   360,9(00)
        + 2(0) × 4 = 8(0)
                   360,98(0)
       + 1 × 8(0)  = 8(0)
                   361,06(0)
```

Remember: **361,060**

Finally, we multiply 2(0) × 8(0), for 1,6(00), and that brings the base number to **362,660.** Then multiply 1 × 4, giving us 4, and add that to the base number to get **the answer: 362,664.**

```
6 2 1      621    THINK: 2(0) × 8(0) = 1,6(00)
  ↓ ↓    × 584                    + 361,06(0)
5 8 4                               362,66(0)
                                  + 1 × 4 = 4
                                    362,664
```

The answer is **362,664**

Now You Can Do It

Try the strategies for cross-multiplication on these numbers. When you get to the third example, remember that the product of 3 × 4(0) is 12(0).

$$\begin{array}{r} 24 \\ \times\ 12 \\ \hline \end{array} \quad \begin{array}{r} 28 \\ \times\ 23 \\ \hline \end{array} \quad \begin{array}{r} 42 \\ \times\ 3 \\ \hline \end{array} \quad \begin{array}{r} 16 \\ \times\ 14 \\ \hline \end{array}$$

$$\begin{array}{r} 91 \\ \times\ 28 \\ \hline \end{array} \quad \begin{array}{r} 108 \\ \times\ 33 \\ \hline \end{array} \quad \begin{array}{r} 243 \\ \times\ 546 \\ \hline \end{array} \quad \begin{array}{r} 2629 \\ \times\ 72 \\ \hline \end{array}$$

USING THE FIVE KEYS ON MULTIPLICATION

The keys to mental math are just as important here as with addition.

Understand what numbers mean. The only way you can keep track of multiplication in your head is to have a firm grip on which digits are units, which are tens, which are hundreds, and so forth. If you can picture this concept and hold it in your mind, you can do even the more difficult right-to-left cross-multiplication mentally.

Think about numbers forward instead of backward. Left-to-right cross-multiplication lets us do this. Some of us, particularly people who work best with words instead of pictures or numbers, find the left-to-right method much easier.

Develop your memory. Obviously, learning the "times" tables lets you figure a lot faster and easier than adding the long way. And if you train yourself to follow a base number in your head, you will soon find you don't need a pencil to multiply or to add.

Practice. Cross-multiplying is a little tricky until you get the technique down pat. Keep at it, though. Practice takes the trick out of the method.

Be creative. Multiplication is full of possibilities. Everywhere you look, there are shortcuts. Look at the following, for example:

Shortcuts and Helpful Hints

When we started to talk about left-to-right cross-multiplication, we found that we could get the answer to 6(00) × 5(00) by adding four zeros to 30 (which is 6 × 5). That's a clue to one quick math shortcut.

If you want to multiply any number by 10, you just add a zero.

1	10	68	239	4,000	98,765
× 10	× 10	× 10	× 10	× 10	× 10
10	**100**	**680**	**2,390**	**40,000**	**987,650**

Something like that happens whenever you multiply by any of the tens—20, 30, 40, 50, 60, 70, 80, 90. Multiply the number by the digit in the tens column, and then put a zero in the ones column of the answer:

10 × 40: 10 × 4 = 40; stick on a zero for **400.**

21 × 40: 21 × 4 = 84; and a zero makes **840.**

The principle applies to all multipliers that end in zero. When you multiply by 100, you tack on two zeros; by 1,000, three zeros. And so on.

We multiplied 5(00) × 6(00):

$$5 \times 6 = 30$$

The five and the six each represented a hundreds number. That's two hundreds. So multiply 30 by a hundred, twice:

$$30 \times 100 = 3{,}000$$

$$3{,}000 \times 100 = \mathbf{300{,}000}$$

When you do that, you get 30 with four zeros.

Or you could arrive at the same answer by multiplying:

$$500 \times 6 = 3{,}000$$

Now only one hundred is left. So you multiply:

$$3{,}000 \times 100 = \mathbf{300{,}000}$$

Look closely and you'll notice something. In each case, the number of magic zeros in the final product is the same as the number of zeros found in both the multiplicand and multiplier combined. Count the zeros in 3,000 × 100: five. Count them in the answer: five!

Count them in 600 × 500: four. In the answer, 30(0,000), there are really four of our magical zeros. We multiplied 6 × 5 for a base number of **30**, then added four zeros for the answer. The first zero after the 3 doesn't count as a magic zero, because it's part of the base number, 30.

Now You Can Do It

Use the keys to human calculating to help find the answers to these challenges:

240	1,970	48	1,070	234	799,600	3,550
× 20	× 10	× 30	× 40	× 34	× 4,000	× 500

4 Complementary Multiplication

HERE'S A THREE-STEP strategy that is mind-boggling in its simplicity. Suppose you want to multiply a couple of tricky-looking numbers—say, 96 and 94.

The first thing you notice is that both numbers are pretty close to 100. Quickly figure out the difference between each number and 100.

The difference between 96 and 100 is 4.

The difference between 94 and 100 is 6.

Let's write these numbers down like this:

$$\begin{array}{rr} 96 & 4 \\ \times\ 94 & \underline{6} \\ \hline \end{array}$$

Now, the second step is to *subtract diagonally.* By that I mean subtract 6 from 96, or 4 from 94. 96 − 6 = **90.** Write that down under the first two numbers. (Note that this is the same as 94 minus 4. It does not matter which you choose. Whichever you diagonally subtract, you get the same number.)

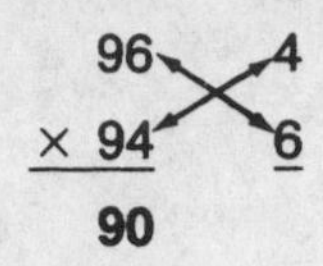

This gives us a base number of 90. That number stands for 90 hundreds, or **90**(00).

In the third step, we multiply the two numbers we got by figuring the difference between our original numbers and 100: 4 and 6. 4 × 6 = **24**. Write that to the right of the 90.

96	4
× 94	× 6
90	**24**

Got that? Guess what? You've got the answer: **9,024**.

Too easy? Let's try it again.

95	5	The difference between 95 and 100 is 5.
× 97	3	The difference between 97 and 100 is 3.
92	**15**	95 − 3 = **92**; 3 × 5 = **15**
		The answer is **9,215.**

If the numbers are larger than 100, you add the difference instead of subtracting. Let's multiply 103 times 107.

The difference between 103 and 100 is three.

The difference between 107 and 100 is seven.

Write the numbers down as before:

```
  103     3
× 107     7
-----
```

Now, instead of subtracting, we will *add diagonally*. You can add 3 plus 7 *or* 7 plus 3; obviously, the result is the same, **110.** Write it under the original numbers.

```
  103 ↖ ↗ 3
× 107 ↙ ↘ 7
-----
  110
```

And now multiply 3 times 7 and write the result, **21,** to the right of the 100.

```
  103       3
× 107     × 7
-----     ---
  110      21
```

There's **the answer: 11,021.**

Here's another example.

```
  106 ↖ ↗ 6     The difference between 106 and 100 is 6.
× 111 ↙ ↘ 11    The difference between 111 and 100 is 11.
-----    --
  117    66     106 + 11 = 117; 6 × 11 = 66.
```

The answer is **11,766.**

Sometimes you will find that you have to carry a digit. For example:

88	13	The difference between 100 and 88 is 12.
× 87	12	The difference between 100 and 87 is 13.
	75	88 − 13 = 75; 12 × 13 is 156.

Now, *don't* add these the conventional way. That is, *don't do this:*

$$\begin{array}{r} 75 \\ +\ 156 \\ \hline \end{array}$$

Instead, we combine these two numbers so that the 1 in the hundreds column in 156 is added to the 5 in 75, which is really in the hundreds column of the answer. Remember, the 75 is really 75(00). So, add it like this:

$$\begin{array}{r} 75\ (00) \\ +\ 1\ \ 56 \\ \hline \mathbf{76\ \ 56} \end{array} \qquad \text{or, if it's easier:} \qquad \begin{array}{l} \ \ \ 75 \\ +\ 156 \\ \hline \mathbf{7656} \end{array}$$

Complementary multiplication works most simply on numbers around 100. However, you can use it on numbers around 50 and numbers around 75, and also on numbers that are near multiples of 100.

Now You Can Do It

Now is a good time to practice complementary multiplication. Even though it's very different from what you may have learned in school, many people find it easy and extremely fast, once they get used to the method. Try these examples:

$$\begin{array}{r} 90 \\ \times\ 93 \\ \hline \end{array} \quad \begin{array}{r} 97 \\ \times\ 94 \\ \hline \end{array} \quad \begin{array}{r} 92 \\ \times\ 96 \\ \hline \end{array} \quad \begin{array}{r} 95 \\ \times\ 99 \\ \hline \end{array} \quad \begin{array}{r} 96 \\ \times\ 85 \\ \hline \end{array}$$

$$\begin{array}{r} 88 \\ \times\ 87 \\ \hline \end{array} \quad \begin{array}{r} 102 \\ \times\ 104 \\ \hline \end{array} \quad \begin{array}{r} 112 \\ \times\ 109 \\ \hline \end{array} \quad \begin{array}{r} 107 \\ \times\ 115 \\ \hline \end{array}$$

MULTIPLYING NUMBERS AROUND 50

First, I want you to keep a fact in mind: 50 is half of 100.

Remember the first part of the answer to 94 × 96, which we got by subtracting diagonally? It was 90, and we said it was a base number. Well, the last digit of that base number always lands in the hundreds column. So when we got **90** by subtracting 4 from 94 or 6 from 96, it was really **90**(00). That's why we tacked on the second half of the answer, instead of adding it as 94 + 24 = 118 (a wrong answer). We were actually adding 90(00) and 24:

$$\begin{array}{r} \mathbf{90}(00) \\ +\ 24 \\ \hline \end{array}$$

90 24, or **9,024**

Now, with numbers near 50 instead of numbers near 100, the first part of the answer still falls in the hundreds column. But because 50 is half of 100, the base number—which forms the first part of our answer—will be half what we get when we subtract diagonally.

Let's multiply 46 times 44. Both numbers are close to 50. Figure the difference between 50 and each number, and write those numbers down in the same way we did for 96 × 94.

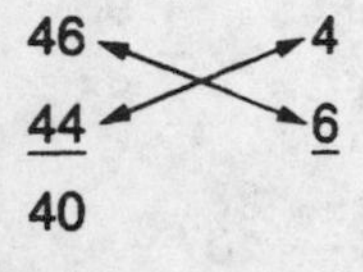

The difference between 50 and 46 is 4.

The difference between 50 and 44 is 6.

Subtract diagonally. The answer is the same: 44 − 4 = 40; 46 − 6 = 40.

Now, 40 is *not* the first part of the answer. Because 50 is half of 100, you take one-half of 40 to get a base number, which is the start of your answer.

40 ÷ 2 = **20** 40 divided by 2 is 20, a base number.

Just as you did with numbers near 100, multiply the two differences together to get the last part of the answer.

4 × 6 = **24**

Remember that our base number of 20 is really 20 hundreds, and add the two together:

20(00)
+ **24**
20 24 **The answer** is **2,024.**

Let's run through that again: What is 48 × 42?

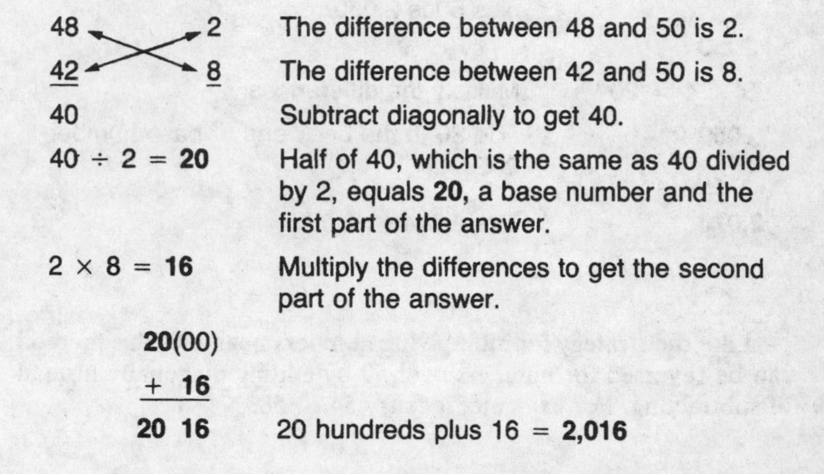

In case you thought that was too simple, there's a trick: If the result of subtracting diagonally is an odd number, you're going to end up with a fraction when you divide it in half. What to do?

It's simple: Remember that *the first part of the answer—the base number—is really in the hundreds*! Half of a hundred is 50. One quarter of a hundred is 25. Three quarters of a hundred is 75. So, if you get, say, a 45 after you subtract diagonally, you're going to end up with 22½ when you divide it in half. That's 22(00) plus half of a hundred, or 50, or 2,250.

We'd better try that on some real numbers. Suppose we were multiplying 45 × 46.

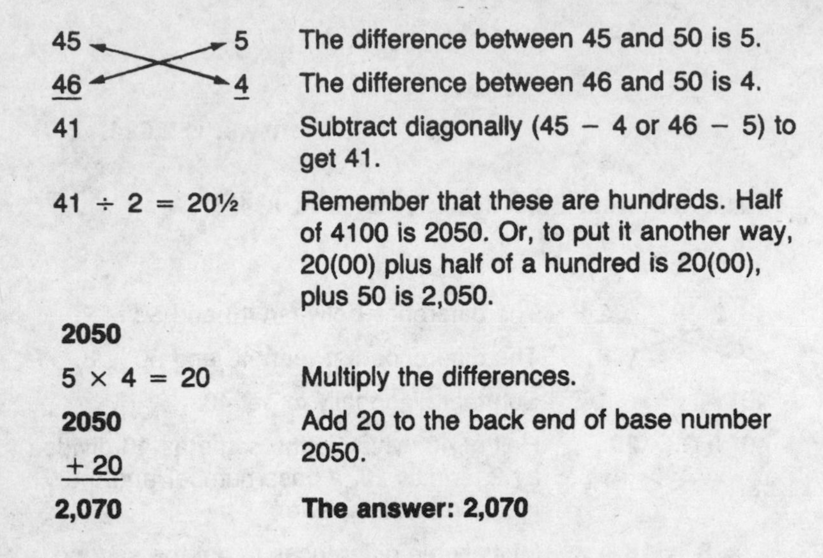

45 ↘↗ 5	The difference between 45 and 50 is 5.
46 ↗↘ 4	The difference between 46 and 50 is 4.
41	Subtract diagonally (45 − 4 or 46 − 5) to get 41.
41 ÷ 2 = 20½	Remember that these are hundreds. Half of 4100 is 2050. Or, to put it another way, 20(00) plus half of a hundred is 20(00), plus 50 is 2,050.
2050	
5 × 4 = 20	Multiply the differences.
2050 + 20	Add 20 to the back end of base number 2050.
2,070	**The answer: 2,070**

Like the strategy for multiplying numbers near 100, this method can be reversed for numbers over 50 by adding diagonally instead of subtracting. For example, let's try 54 × 58.

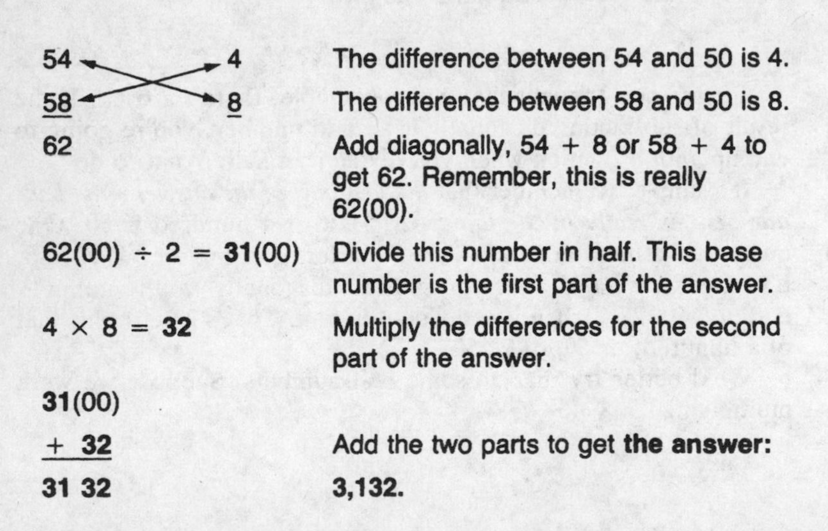

54 ↘↗ 4	The difference between 54 and 50 is 4.
58 ↗↘ 8	The difference between 58 and 50 is 8.
62	Add diagonally, 54 + 8 or 58 + 4 to get 62. Remember, this is really 62(00).
62(00) ÷ 2 = **31**(00)	Divide this number in half. This base number is the first part of the answer.
4 × 8 = **32**	Multiply the differences for the second part of the answer.
31(00) + **32**	Add the two parts to get **the answer:**
31 32	**3,132.**

Here again, if the result of adding is an odd number, you will end up with a fraction when you divide in half. Express that as a part of 100—50 for ½, 25 for ¼, or 75 for ¾.

Now You Can Do It

Once you understand that your base number is really in the hundreds, it's easy to write the fractions that sometimes come from dividing as digits in the tens. Try complementary multiplication on these examples:

$$\begin{array}{r} 45 \\ \times 49 \\ \hline \end{array} \quad \begin{array}{r} 48 \\ \times 44 \\ \hline \end{array} \quad \begin{array}{r} 40 \\ \times 47 \\ \hline \end{array} \quad \begin{array}{r} 37 \\ \times 42 \\ \hline \end{array}$$

$$\begin{array}{r} 54 \\ \times 55 \\ \hline \end{array} \quad \begin{array}{r} 52 \\ \times 57 \\ \hline \end{array} \quad \begin{array}{r} 58 \\ \times 60 \\ \hline \end{array} \quad \begin{array}{r} 61 \\ \times 53 \\ \hline \end{array}$$

MULTIPLYING NUMBERS AROUND 25

A similar strategy works on numbers near 25. In this case, we bear in mind that 25 is one fourth of 100. When we get the result of subtracting diagonally, therefore, we will want a quarter of it, instead of a half. How do you find a quarter of a number? Divide by four.

Let's multiply 21 × 24.

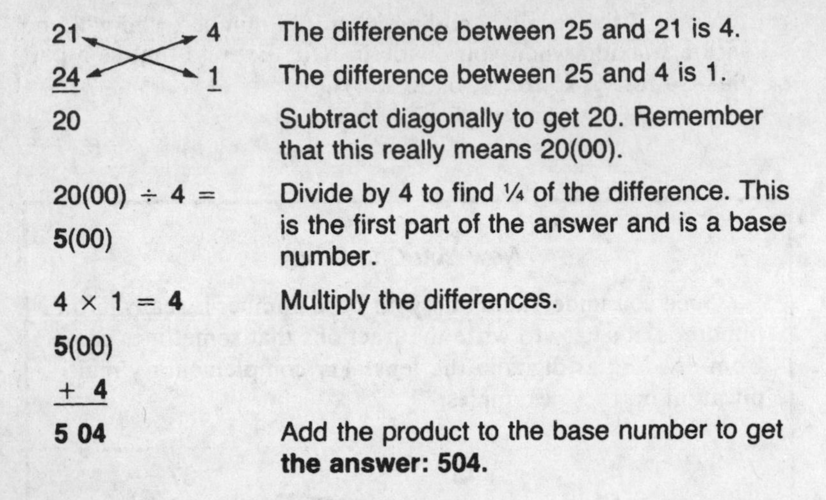

21 ⟶ 4 24 ⟶ 1	The difference between 25 and 21 is 4. The difference between 25 and 4 is 1.
20	Subtract diagonally to get 20. Remember that this really means 20(00).
$20(00) \div 4 =$ **5**(00)	Divide by 4 to find ¼ of the difference. This is the first part of the answer and is a base number.
$4 \times 1 =$ **4**	Multiply the differences.
5(00) + **4** **5 04**	Add the product to the base number to get **the answer: 504.**

Does it work on numbers over 25, if we add instead of subtracting diagonally? Sure it does! Let's try it on 26×32.

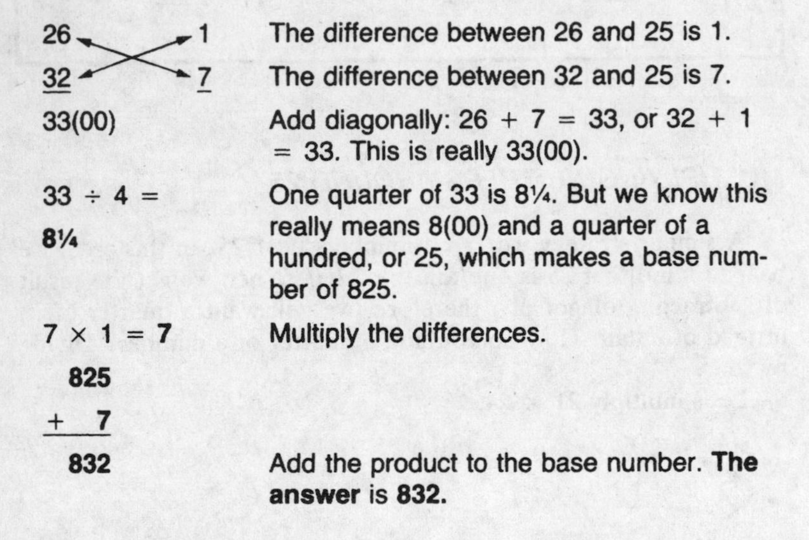

26 ⟶ 1 32 ⟶ 7	The difference between 26 and 25 is 1. The difference between 32 and 25 is 7.
33(00)	Add diagonally: $26 + 7 = 33$, or $32 + 1 = 33$. This is really 33(00).
$33 \div 4 =$ **8¼**	One quarter of 33 is 8¼. But we know this really means 8(00) and a quarter of a hundred, or 25, which makes a base number of 825.
$7 \times 1 =$ **7**	Multiply the differences.
825 + **7** **832**	Add the product to the base number. **The answer** is **832.**

Now You Can Do It

Practice is like physical exercise—it develops the muscles of the mind. Work out on a few of these muscle machines:

26 ×28	24 ×21	28 ×22	22 ×23	25 ×29
33 ×24	37 ×32	37 ×27	34 ×39	

MULTIPLYING NUMBERS AROUND 75

To use the strategy on numbers near 75, you have to be in top *mathletic* shape, because it involves another step. It can be done. The principle is exactly the same. This time, though, instead of taking a half or a quarter of the result of your diagonal figuring, you need to find three quarters of the result to find the base number. To do that, you divide the number by 4 and then multiply the answer by 3. Take 70 × 73:

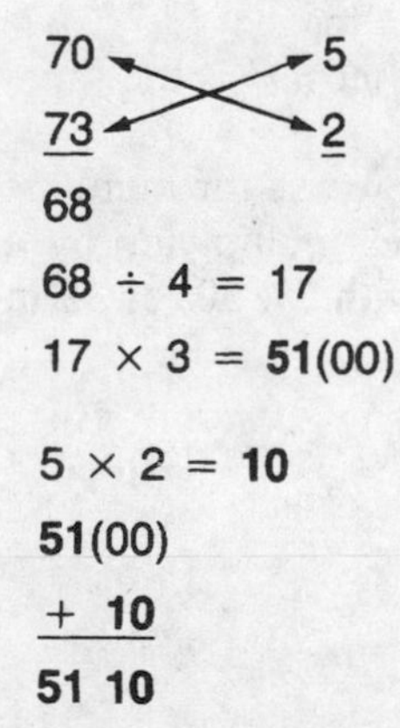

The difference between 75 and 70 is 5.

The difference between 75 and 73 is 2.

Subtract diagonally to get 68.

To get ¾ of 68, first divide by 4; then multiply by 3. Remember that this base number is really in the hundreds.

Multiply the differences.

Add the product to the base number for **the answer: 5,110.**

The added step of multiplying by 3 complicates things when you run into fractions. Take 71 × 74:

71 ↘↗ 4	The difference between 75 and 71 is 4.
74 ↗↘ 1	The difference between 75 and 74 is 1.
70(00)	Subtract diagonally to get the difference.
70 ÷ 4 = 17½	Divide by 4; this gives you ¼.
17½ × 3 = **52½**	Multiply this by 3 to get ¾. We know that 52 is actually 52(00), and so the ½ is actually half of a hundred, which is the same as 50. We have a base number of 5,250.
4 × 1 = **4**	Multiply the differences.
5,250 + **4** = **5,254**	Add the product to the base number for **the answer: 5,254.**

Even though this is a little more involved than multiplying numbers around 25, 50, or 100, with practice it's not difficult. That's why *practice* is one of the five keys to human calculating!

Remember: When you're dealing with hundreds, ¼ is the same as 25; ½ is 50; and ¾ is 75.

Here, as with 25, 50, and 100, the strategy works for numbers higher than 75, too. Because one step involves multiplying by a fraction, you almost always will have to deal with 25, 50, or 75 in your base number. Let's multiply 76 × 80.

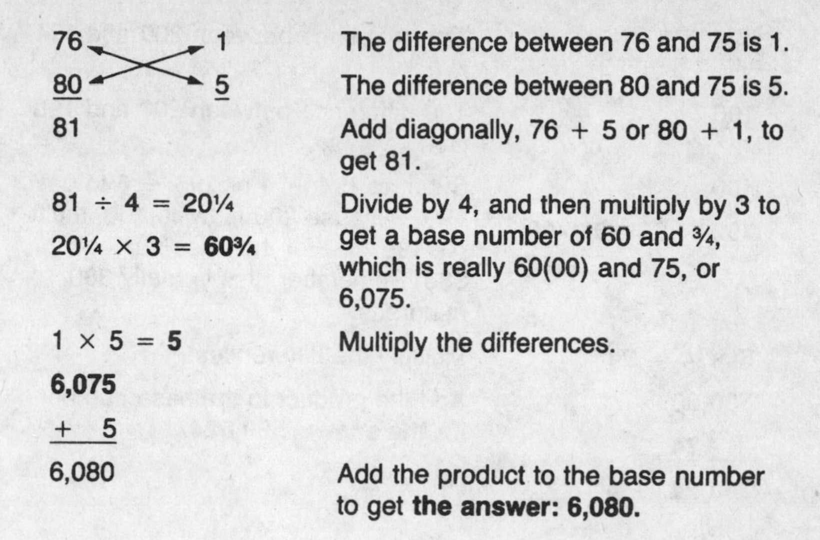

76 ⟶ 1 80 ⟶ 5	The difference between 76 and 75 is 1. The difference between 80 and 75 is 5.
81	Add diagonally, 76 + 5 or 80 + 1, to get 81.
$81 \div 4 = 20\frac{1}{4}$ $20\frac{1}{4} \times 3 = \mathbf{60\frac{3}{4}}$	Divide by 4, and then multiply by 3 to get a base number of 60 and ¾, which is really 60(00) and 75, or 6,075.
$1 \times 5 = \mathbf{5}$	Multiply the differences.
6,075 + 5 6,080	Add the product to the base number to get **the answer: 6,080.**

Now You Can Do It

Ready for the 10K run in complementary multiplication? This could be a serious workout!

$$\begin{array}{r} 71 \\ \times\ 74 \\ \hline \end{array} \quad \begin{array}{r} 73 \\ \times\ 62 \\ \hline \end{array} \quad \begin{array}{r} 75 \\ \times\ 72 \\ \hline \end{array} \quad \begin{array}{r} 78 \\ \times\ 79 \\ \hline \end{array} \quad \begin{array}{r} 68 \\ \times\ 73 \\ \hline \end{array}$$

$$\begin{array}{r} 70 \\ \times\ 67 \\ \hline \end{array} \quad \begin{array}{r} 82 \\ \times\ 78 \\ \hline \end{array} \quad \begin{array}{r} 79 \\ \times\ 80 \\ \hline \end{array} \quad \begin{array}{r} 64 \\ \times\ 74 \\ \hline \end{array}$$

MULTIPLYING WITH NUMBERS NEAR MULTIPLES OF 100

Just as this strategy works easily when you're dealing with numbers on either side of 100, it works for numbers around 200, 300, 400, etc. To get your base number, you multiply the result of diagonal adding or subtracting by 2, 2, 4, etc. For example, 194 × 196:

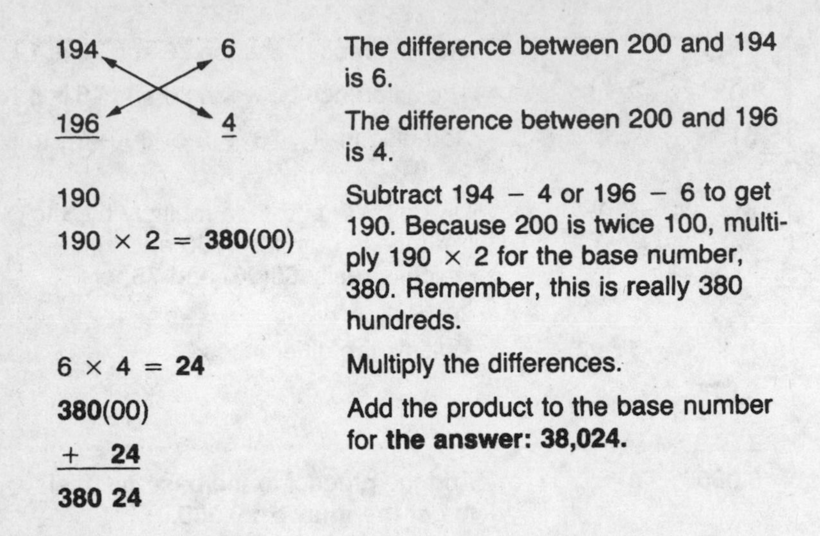

194 6	The difference between 200 and 194 is 6.
196 4	The difference between 200 and 196 is 4.
190 190 × 2 = **380**(00)	Subtract 194 − 4 or 196 − 6 to get 190. Because 200 is twice 100, multiply 190 × 2 for the base number, 380. Remember, this is really 380 hundreds.
6 × 4 = **24**	Multiply the differences.
380(00) + **24** **380 24**	Add the product to the base number for **the answer: 38,024.**

Yes, it works on numbers larger than a multiple of 100, too. Let's try 411 × 422:

411 11	The difference between 411 and 400 is 11.
422 22	The difference between 422 and 400 is 22.
433	Adding diagonally, 411 + 22 or 422 + 11 = 433.
433 × 4 = **1,732**	Multiply by 4, because 400 is 4 × 100. This base number is really 1,732 hundreds, or 1732(00).
11 × 22 = **242** **1732**(00) + **2 42**	Multiply the differences. Add the product to the base number.
1734 42	**The answer** is **173,442.**

Now You Can Do It

After the workout on numbers near 75, these mental gymnastics should be a piece of cake. Try them on your friends!

$$\begin{array}{r} 221 \\ \times\ 215 \\ \hline \end{array} \quad \begin{array}{r} 297 \\ \times\ 292 \\ \hline \end{array} \quad \begin{array}{r} 382 \\ \times\ 394 \\ \hline \end{array} \quad \begin{array}{r} 495 \\ \times\ 488 \\ \hline \end{array}$$

$$\begin{array}{r} 512 \\ \times\ 520 \\ \hline \end{array} \quad \begin{array}{r} 698 \\ \times\ 691 \\ \hline \end{array} \quad \begin{array}{r} 995 \\ \times\ 998 \\ \hline \end{array}$$

SHORTCUTS AND HELPFUL HINTS

You can instantly multiply any number by 101 just by multiplying the number by 100 and adding the number.

$$18 \times 101 = \mathbf{1{,}818}$$
$$84 \times 101 = \mathbf{8{,}484}$$
$$97 \times 101 = \mathbf{9{,}797}$$

Why does this work? Remember that multiplying is really a form of adding. When we say "18 × 101," we're really saying the same as "18 + 18 + 18 . . . 101 times. If we go "18 × 100," we've added 18 one hundred times. So to multiply 18 times 101—which is the same as adding 18 to itself 101 times—all we have to do is multiply it by 100 (which is easy) and then add one more 18 to the product. You could think of the examples above like this:

$$18 \times 101 = (18 \times 100) + (18) = 1{,}800 + 18 = \mathbf{1{,}818}$$
$$84 \times 101 = (84 \times 100) + (84) = 8{,}400 + 84 = \mathbf{8{,}484}$$
$$97 \times 101 = (97 \times 100) + (97) = 9{,}700 + 97 = \mathbf{9{,}797}$$

To multiply a number by 99, multiply it by 100 and then subtract the number. To figure the answer to 99 × 42, for example:

$$\begin{array}{r} 42 \\ \times \quad 100 \\ \hline 4{,}200 \\ - \quad 42 \\ \hline \mathbf{4{,}158} \end{array}$$

The same principle is at work here. Multiplying a number by 99 is the same as adding it 99 times. We know it's easy to multiply by 100. In this case, though, 99 is one *less than* 100, and so we *subtract* the number once from the product. You could think of the example above like this:

$$42 \times 99 = (42 \times 100) - 42 = 4{,}200 - 42 = \mathbf{4{,}158.}$$

Okay. So how would you multiply a number by 98?

We see that 98 is two less than 100, and we know that multiplying the number by 98 is the same as adding it 98 times. Multiplying the number by 100 is the same as adding it 100 times. To find the product of the number times 100, first multiply it by 100. Double the number and then subtract the result from the product of the number times 100. For 3 × 98, for example:

$$\begin{array}{r} 3 \\ \times\ 100 \\ \hline 300 \\ - \quad 6 \\ \hline \mathbf{294} \end{array} \quad \text{THINK: } 3 \times 2 = 6$$

And picture the process in this way:

$$3 \times 98 = (3 \times 100) - (3 \times 2) = 300 - 6 = \mathbf{294}$$

Or, if you prefer:

$$3 \times 98 = (3 \times 100) - 3 - 3 = 300 - 6 = \mathbf{294}$$

Every time you are faced with a difficult-seeming problem, look at it to see if the numbers are close to something easy. You may find it surprisingly simple. Remember, *creativity* is one of the five keys, and a very important one at that!

5 Box Multiplication

HERE'S A STRATEGY THAT appeals to "word people"—people who are good in English but think they can't do math. Language buffs like this method because it reminds them of a crossword puzzle. It's also an excellent tool for kids who have trouble keeping numbers organized, because it gets the answer while it helps you keep your figures in order. "Box multiplication," as I call it, works by breaking a challenge down into simpler problems.

Suppose you want to multiply 24 by 12. First, draw a box and divide it into four squares, one for each of the four digits in the problem. Write the digits next to the squares in the order that they appear. Put the digits of the first number along the top of the box, starting with the top left square. Put the digits of the second number along the right side of the box, starting with the top right square and moving down the side. Now draw lines to divide the four little boxes in half diagonally:

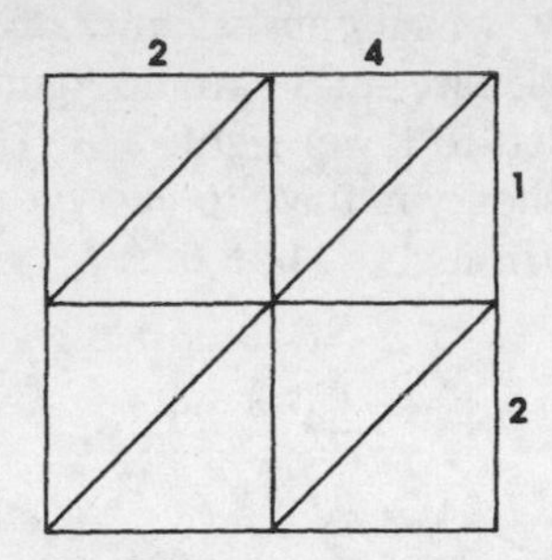

Next, you multiply around the box, entering your answer in the top and bottom halves of the boxes you have made. The units digits go into the bottom parts, and tens go into the top parts. Starting at the top left, $2 \times 1 = 2$. That's a unit with no tens, so you put it in the lower triangle of the 2×1 box. Drop down to the box below it: $2 \times 2 = 4$. Put the 4 in the lower triangle of that box. Move over to the 4 in 24, and multiply $4 \times 1 = 4$; put it in the lower triangle of its box. And last, $4 \times 2 = 8$, which also goes in its lower triangle.

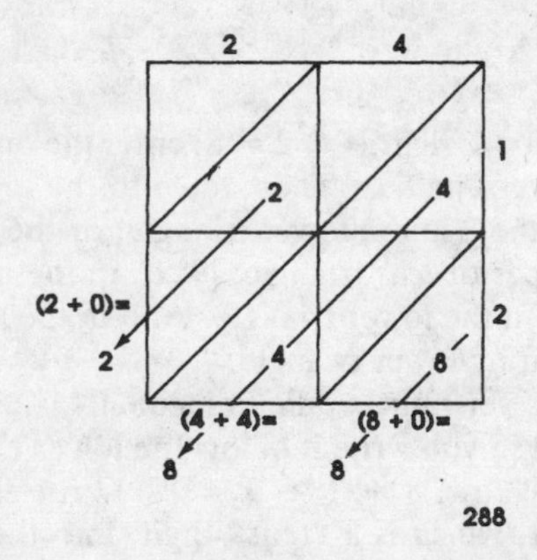

Now we are going to add the numbers in the box diagonally, starting with the lower right-hand box. There's nothing to add to 8, so just write that below its box. In the next diagonal row up, starting with the lower triangle in the square on the upper right, the 4, add diagonally down and to the left. In this case, $4 + 4 = 8$. Write that beneath its box. The only number in the next row, which starts with the upper triangle on the top right box, is a 2. Write it next to the box. In the last row, which consists only of the upper triangle in the upper left-hand square, there are no figures. Now, read off the figures from upper left to lower right: **288**. That's the answer.

What happens when you have to carry a number? You carry it to the next segment in its box, like this:

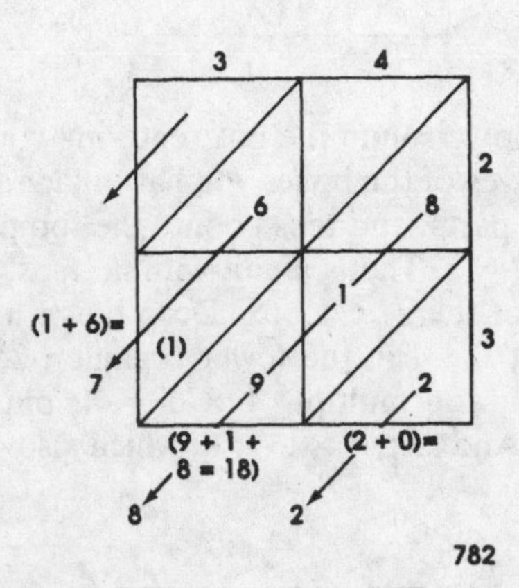

Starting on the upper right, 4 × 2 = 8; enter the one-digit answer in the lower triangle of its box. Then down to the lower right box, 4 × 3 = 12; enter the 2 in the lower triangle, or the *units* triangle, and the 1 in the upper or *tens* triangle. Next, in the upper left box, 3 × 2 = 6; place it in the lower triangle. And in the lower left box, 3 × 3 = 9; put it in its lower triangle.

Start from the lower right, adding diagonally: the lower right 2 stands by itself, and so you write it below the lower right-hand box. In the next diagonal row, 8 + 1 + 9 = 18. Write the 8 below its row and carry the 1, which is a "tens" digit, into the tens triangle of its box, putting it in the next row to be diagonally added.

Now add that row: 1 + 6 = 7. Write it on the left side of the box next to its row. All that remains is to read **the answer: 782.**

This works with numbers of any size. You just draw larger boxes. Visualize 345 × 27 as it appears below.

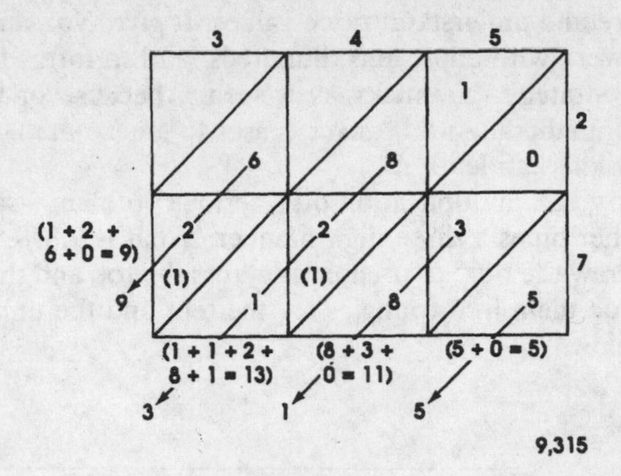

THINK: 3, 4, and 5 go across the top.

2 and 7 go down the right side.

$5 \times 2 = 10$; write 1 in the tens triangle and 0 in the units triangle.

$5 \times 7 = 35$; 3 in the tens and 5 in the units triangle.

$4 \times 2 = 8$; into the units triangle.

$4 \times 7 = 28$; 2 in the tens and 8 in the units triangle.

$3 \times 2 = 6$; 6 in the units triangle.

$3 \times 7 = 21$; 2 in the tens and 1 in the units triangle.

Add diagonally from the lower right:

5 alone is 5.

$8 + 3 + 0 = 11$; write 1 and carry 1 to the next row.

The carried $1 + 1 + 2 + 8 + 1 = 13$; write 3 and carry 1 to the next row.

The carried $1 + 2 + 6 = 9$; write it by its row.

There are no figures in the last row.

Read **the answer: 9,315.**

As you can see, one of this method's beauties is that it helps you picture and understand place values. It gives you a diagram in which to work with ones, tens, hundreds, and so forth. I've always liked this strategy, gimmicky as it seems, because of the way it organizes numbers. And it never ceases to amaze me how quickly and easily kids can learn it.

Let's try box multiplication on a serious problem—say, a four-digit number times a three-digit number. What is 9,878 × 849?

First draw the box: four cubicles across the top and three on the side. Divide them into triangles for the tens and the units, as they appear below.

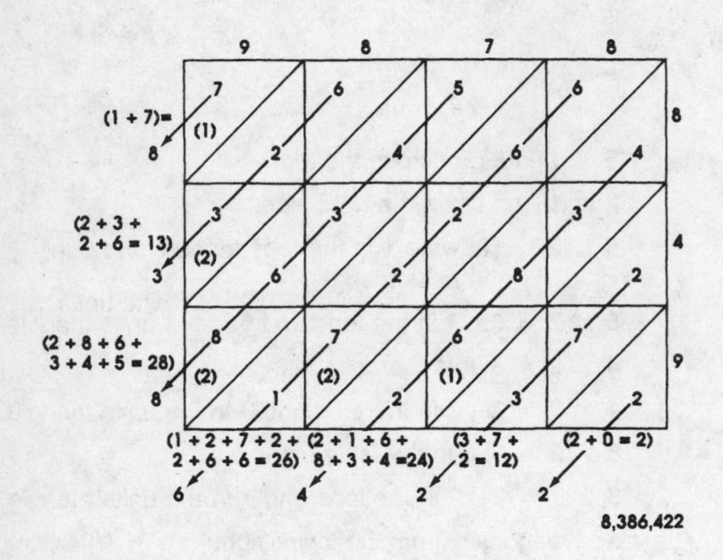

THINK: 9, 8, 7, and 8 go across the top.

8, 4, and 9 go down the right side.

8 × 8 = 64; put the 6 in the tens triangle and the 4 in the units triangle.

8 × 4 = 32; write the 3 in the tens triangle and the 2 in the units triangle.

8 × 9 = 72; 7 in the tens triangle and 2 in the units triangle.

Moving left along the top row:

7 × 8 = 56; 5 in the tens triangle and 6 in the units triangle.

7 × 4 = 28; 2 in the tens triangle, 8 in the units triangle.

7 × 9 = 63; 6 in the tens triangle, 8 in the units triangle.

Move left again:

8 × 8 = 64; 6 to the tens triangle, and 4 to the units triangle.

8 × 4 = 32; 3 in the tens triangle; 2 in the units triangle.

8 × 9 = 72; 7 in the tens and 2 in the units.

And finally, 9 × 8 = 72; 7 in the tens triangle and 2 in the units triangle.

9 × 4 = 36; 3 in the tens triangle and 6 in the units triangle.

9 × 9 = 81; 8 in the tens triangle, 1 in the units triangle.

Add diagonally, starting at the lower right:

2 alone is 2.

3 + 7 + 2 = 12; write down the 2 and carry the 1(0) to the next row.

2 + 1 + 6 + 8 + 3 + 4 = 24. Write down the 4 and carry the 2.

1 + 2 + 7 + 2 + 2 + 6 + 6 = 26. Repeat what you did in the last step.

2 + 8 + 6 + 3 + 4 + 5 = 28; write down the 1 and carry the 2.

2 + 3 + 2 + 6 = 13; write down the 3 and carry the 1.

1 + 7 = 8.

Now read off **the answer: 8,386,422.**

Those of us who learned to multiply the traditional way—by writing the product of each digit in offset lines, one beneath the other, and then adding, need only think a moment to see that box multiplication does essentially the same thing. Visualize 48 × 16 in each strategy.

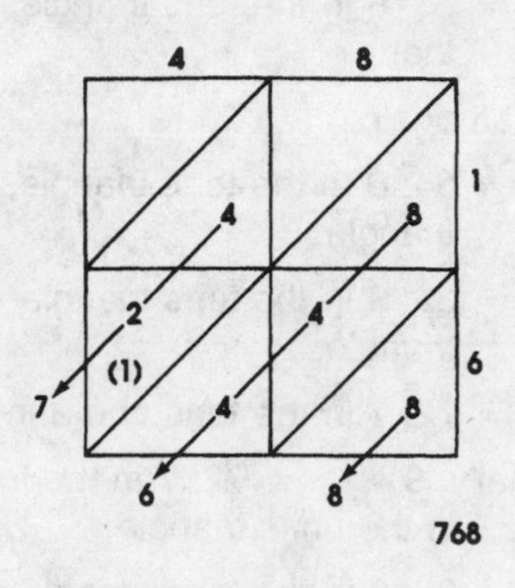

$$
\begin{array}{r}
48 \\
\times\ 16 \\
\hline
(1)\ 288 \\
48 \\
\hline
768
\end{array}
$$

Remember how hard it was, unless you were the tidy type, to keep those columns straight? Just think of the erasures, scribbles, and mistakes you made when you tried to multiply two figures whose product came to more than 6 million! Box multiplication lets you add up each digit's product without having to write down a series of long numbers in visually confusing form.

Now You Can Do It

To use this strategy, you bring two of the five keys to human calculating into play: memory and practice. Once you have learned the times table, box multiplication becomes extremely easy. You learn your times table with lots of practice. Try these challenges:

$$\begin{array}{r} 28 \\ \times\ 14 \\ \hline \end{array} \qquad \begin{array}{r} 312 \\ \times\ 33 \\ \hline \end{array} \qquad \begin{array}{r} 567 \\ \times\ 98 \\ \hline \end{array} \qquad \begin{array}{r} 453 \\ \times\ 286 \\ \hline \end{array} \qquad \begin{array}{r} 2{,}345 \\ \times\ 64 \\ \hline \end{array} \qquad \begin{array}{r} 5{,}745 \\ \times\ 379 \\ \hline \end{array}$$

6 Squaring

SQUARING IS MULTIPLYING A number by itself. For example, $2 \times 2 = 4$, or $2{,}489 \times 2{,}489 = 6{,}195{,}121$. The answer—in these examples, 4 and 6,195,121—is said to be the *square* of the number. Or we may say that 6,195,121 is *2,489 squared,* and we write that expression like this: $2{,}489^2$. So, we can write about squaring and its results in this way:

$$2^2 = 4$$

$$4^2 = 16$$

$$5^2 = 25$$

$$10^2 = 100$$

$$2{,}489^2 = 6{,}195{,}121$$

Once you've learned the times table, it's easy to square the numbers from one to 12—you already know them. When you get into larger numbers, such as 64, you start to meet some challenges.

There are ways to simplify squaring. Here are three strategies that will make you look like a real mathlete.

SQUARING NUMBERS AROUND 100

If a number falls under 100 and above 51, you simply figure the difference between 100 and the number. Subtract the difference from the number being squared, and write the answer as the first digits in your answer.

Let's try this on the number 96. We see right away that 96 is 4 less than 100. Then we subtract this difference from the number being squared:

$$\begin{array}{r} 96^2 \\ -\ 4 \\ \hline 92 \end{array}$$

Subtract that amount from the number.

The first two digits of this number (92) form the first part of your answer. I want you to keep in mind that the last digit of this number is *actually in the hundreds column of the answer.* Visualize it as 92(00).

Now, to get the second part of the answer, take the difference between 100 and the number being squared (in this case, 4), and square it:

$$4^2 = 16$$

Place that number after the digits you got in your first step, and you have the answer: **9,216.**

$$96^2 = \mathbf{9{,}216}$$

Let's run through that again with a number in the 90s. What's the square of 92?

92^2	**THINK:** The difference between 100 and 92 is 8.
$-\ 8$	Subtract 8 from 92.
84(00)	This is the first part of the answer. Remember that the 4 is in the hundreds column.
8^2 = **64**	**THINK:** Square the difference between 92 and 100. This is the last part of the answer.
92^2 = **8,464**	Put the two together to see **the answer.**

Numbers less than 91 create a new problem: The square of the difference is larger than 99. If the squared difference is larger than 99, add the digit in the hundreds column to the second digit in the first part of the answer. For example:

88^2	**THINK:** 88 is 12 less than 100.
$-\ 12$	Subtract 12 from 100.
76	First part of the answer.

Now, remember that this **76** is really **76**(00). *The last digit of the number you get when you subtract the difference-from-100 from the number being squared is always a hundreds digit.*

In the next step, you square the difference between 100 and 88. That difference is 12. Now, 12^2 is 144, and that is larger than 99. Knowing that our **76** is really **76**(00) tells us that we have to break our 144 into 1(44) and add accordingly. We'll add the 1(00) in 144 to the hundreds column in 76(00). Then we'll add the 44 in the tens and units columns.

$12^2 = 144$	Square the difference of 100 − 88. It's more than 99; so, add 1(00) in 144 to the 76(00) in the first part of the answer.
76(00)	
+ 1(00)	
77(00)	Now add the 44 remaining in the 1(44) we got by squaring 12.
77(00)	Remember their place values.
+ 44	
77 44	**The answer: 7,744**

If the number to be squared is more than 100, you *add* the difference to the number, rather than subtracting. Complete the strategy in the same way as you did with numbers from 52 to 100. What is 106^2?

	THINK: The difference between 106 and 100 is 6.
106^2	
+ 6	Add 6 to 106.
112	This is the first part of the answer.
6^2 = **36**	Square the difference between 100 and 106.
11,236	Put the two together to see **the answer.**

Again, if the difference squared has more than two digits, add this hundreds digit to the hundreds digit in the first part of the answer, just as you did before. This will always happen when you square numbers larger than 109. For example:

110^2	**THINK:** The difference between 110 and 100 is 10.
+ 10	Add 10 to 110.
120	The first part of the answer.
$10^2 =$ 100	Square the difference. It has more than two digits—that is, it's larger than 99.
120(00)	Add 100 to 120(00).
+ 1 00	
121 00	**The answer** is **12,100.**

Let's try that with a number whose answer will not end in zero; say, 112^2.

112^2	**THINK:** The difference between 112 and 100 is 12.
+ 12	Add 12 to 112.
124(00)	The first part of the answer.
$12^2 = 144$	Square the difference.
124(00)	Add 144 to 124(00).
1 44	
125 44	**The answer** is **12,544.**

Now You Can Do It

Remember, the first key to human calculating is understanding numbers. When it comes to squaring, that means keeping the place values straight. Just keep in mind that, with numbers near 100, the last digit in the first part of the answer *actually falls in the hundreds column.* When you add the second part, be sure to add units to units, tens to tens, hundreds to hundreds, and so on. Express these numbers in their longer form:

94^2 89^2 113^2 84^2 109^2 127^2

The third key to human calculating is memory. Write the answers to these squares. If you don't already know them like your own name, write each one ten times. Then recite the answers without looking at your crib sheet.

$1^2 =$

$2^2 =$

$3^2 =$

$4^2 =$

$5^2 =$

$6^2 =$

$7^2 =$

$8^2 =$

$9^2 =$

$10^2 =$

$11^2 =$

$12^2 =$

Teaser: If writing each of these expressions takes you two seconds, and writing a math fact ten times commits it to your memory, how many minutes will it take to learn all 12 squares?

SQUARING NUMBERS AROUND 1,000

You apply the same thinking when you square a number that's near 1,000. The trick is to make sure you keep track of your place values. Let's try it on 996^2:

996^2	The difference between 1,000 and 996 is 4.
$996 - 4 = 992$ **992**(000)	Subtract 4 from the number to be squared; this is the first part of the answer: **992**(000). Remember that you're now talking about 992 thousands.
$4^2 = 16$	Square the difference between 996 and 1,000. Remember that this is 16, not
992(000) + 16 **992 016**	16(0). Write the number in the tens and units columns, leaving a zero in the hundreds column. **The answer,** then, is **992,016.**

This strategy works for numbers higher than 1,000, just as it works for numbers above 100.

$1{,}007^2$ + 7 **1,014** Really **1,014**(000) $7^2 = 49$	The difference between 1,007 and 1,000 is 7. Add 7 to 1,007. This is the first part of the answer: **1,014**(000). The final digit lands in the thousands column. Square the difference. The answer falls in the tens and units columns.
+ **1,014**(000)	Combine the two to find . . .
1,014,049	**the answer!**

Now You Can Do It

Don't forget the place values of the digits you come up with.

993^2 998^2 $1{,}004^2$ 998^2 $1{,}006^2$ 976^2 $1{,}012^2$

SQUARING NUMBERS THAT END IN 5

Any number that ends in 5 is easy to square.

To start with, you know that $5 \times 5 = 25$. So the answer will automatically end in 25.

To get the first part of the answer, just take the number's first digit, add 1 to it, and multiply the result by the first digit of the number to be squared. For example:

What is 35^2?

THINK: The answer will end in **25.**

$$\begin{array}{r} 1 \\ +\ 3(5) \\ \hline 4 \end{array}$$

THINK: Add 1 to the first digit; ignore the rest of the number.

$4 \times 3 =$ **12**

THINK: Multiply the result times the first digit of the original number. This result is the first part of the answer.

12 with **25** makes **1,225**

THINK: Stick the first part of the answer onto the 25 to get **the whole answer: 1,225.**

If you're using this technique with a larger number, add 1 to the number that's formed by the digits to the left of the 5. Let's find the square of 105.

105^2	The last part of the answer will be **25.**
1	
+ 10	Add 1 to the digits to the left of the 5.
11	The result is 11. Multiply that times the 10 in the original number.
× 10	
110	This is the first part of the answer.
11,025	The two parts go together for **the answer: 11,025.**

You may have to use pencil and paper to square larger figures, but the strategy is the same and it's much easier than multiplying out the whole problem.

There are plenty of other tricks to make squaring easier.

SQUARING NUMBERS THAT END IN 1

Consider this: Numbers that end in zero, up to 120, are easy to square if you've learned your times table. For example, $120^2 = 120 \times 120 = 12 \times 12$ and two 0s = 144 and 00, or **14,400.** So, when you want to square a number that ends in 1, first square the next lower number—it will always end in zero.

21^2 — THINK: The next lower number is 20.

$20 \times 20 = 400$

Now, add that next lower number (20) and the number being squared (21):

$$20 + 21 = 41$$

Add the two results together to get the answer.

400
+ 41
441

Let's try that on a larger number. How about 81^2?

80	The next lower number is 80.
× 80	
6,400	
80 + 81 = 161	Add the next lower number plus the number.
6,561	

We can apply this principle to numbers that come close to ending in 5, just as we apply it to numbers that almost end in 0.

SQUARING NUMBERS THAT END IN 4

This strategy uses a technique we've already learned, squaring numbers that end in 5. Suppose you want to find the answer to 34^2. Instead of 34, we'll first square the next higher number, 35. The number higher than a number ending in 4 will always end in 5, of course, and we know a fast way to square numbers ending in 5. As we've seen, $35^2 = 1,225$.

Now, add the number being squared (34) to the next higher number (35). The result is 69.

Subtract this from the square of the larger number, and the result will be your answer:

35^2 = 1,225	34	1,225
	+ 35	− 69
	69	**Answer: 1,156**

SQUARING NUMBERS THAT END IN 6

Numbers ending in 6 are just on the high side—instead of the low side—of numbers ending in 5. So we can reverse our ending-in-4 strategy for an easy way to square numbers ending in 6. Let's try it on 36^2.

This time, we square the next *lower* number, which naturally will end in 5. Again, add the number to be squared with the 5-number. And this time you *add* the result from the square of the 5-number.

$35^2 = 1{,}225$	35	1,225
	+ 36	+ 71
	71	**Answer: 1,296**

SQUARING NUMBERS THAT END IN 9

This strategy is similar to squaring numbers ending in 4. You square the next higher number (which will always end in zero); then you add the number to be squared to the next higher number and subtract the result from the square you've found. Find, for example the answer to 79^2.

$80^2 = 6{,}400$	79	6,400
	+ 80	− 159
	159	**Answer: 6,241**

Now You Can Do It

Try out what you've learned on these figures:

15^2	85^2	105^2	345^2	$1{,}235^2$	555^2
31^2	71^2	111^2	51^2	61^2	91^2
184^2	14^2	64^2	84^2	54^2	114^2
56^2	106^2	96^2	76^2	16^2	66^2
19^2	99^2	69^2	109^2	59^2	39^2

Teaser: Your friend has spotted an intriguing display at the grocery store. Someone has stacked boxes of Chinese noodle soup mix, one box deep, 24 boxes high, and 24 boxes wide. Ideally, when your friend gives the box in the center of the bottom row a swift kick (as he is about to do), how many boxes of Chinese noodle soup mix will come crashing to the floor? *Hint:* Visualize! The bottom row is already on the floor.

USING SQUARES TO HELP MULTIPLY OTHER NUMBERS

As the exercise above suggests, squaring has lots of practical applications. One that not many people know is that you can use it to simplify certain kinds of multiplication problems. For example:

Multiplying Two Numbers Whose Difference is 1

Suppose you have two numbers—say, 25 and 24—that are only one number apart. If you square the *larger* number—25, in this case—and then *subtract* it from its square, you get the answer to 25 × 24. If you square the *smaller* number and then *add* it to its square, you will get the same answer! Let's try that:

25	$25^2 = 625$	625
× 24	Subtract 25 from 625	− 25
	Answer:	**600**

Those of us who cheated with a hand-held calculator know that 600 is the correct answer to 24 × 25. What happens if we square the smaller number and add?

24	$24^2 = 576$	576
× 25	Add 24 to 576	+ 24
	The same answer!	**600**

Multiplying Two Numbers Whose Difference Is 2

To work this strategy, you have to know what an *average* is and how to get it. The average of two or more figures is the answer that you get when you add up the figures and divide by the number of figures you added. Say you wanted to know the average of three numbers: 4, 8, and 18. First you would add them up:

$$\begin{array}{r} 4 \\ 8 \\ +\ 18 \\ \hline 30 \end{array}$$

Then you would divide the answer, 30, by the number of figures you added together, 3: 30 divided by 3 = **10**. The average of 4, 8, and 18 is **10**.

We'll use averaging in three of the next four strategies.

To multiply two numbers when they have a difference of 2, square the average of those two numbers and then subtract 1. For example, 17 × 15:

17	**THINK:** 17 + 15 = 32; 32 divided by 2 = 16.
× 15	Square the average: 16^2 = 256.
255	Subtract 1: 256 − 1 = **255.**

Remember that the numbers have to be separated by 2, and only by 2, for this method to work.

Multiplying Two Numbers Whose Difference Is 3

To multiply two numbers when they have a difference of 3, first add 1 to the smaller number. Square the sum. Then subtract 1 from the smaller number and add the result to that square. Let's try it on 24 × 21:

24	**THINK:** add 1 to the *smaller* number.
× 21	21 + 1 = 22. Square the sum: 22^2 = 484.
	Now subtract 1 from the smaller number: 21 − 1 = 20. Add it to the squared sum:
504	20 + 484 = **504.**

You can see how quickly a person who is good at squaring can find this product by thinking of this procedure in a kind of shorthand:

24	
× 21	22^2 = 484; + 20 = **504.**
504	

Multiplying Two Numbers Whose Difference Is 4

This strategy again has you square the average of the two numbers. And this time, you subtract 4 from your square.

	THINK: 46 + 42 = 88; 88 divided by 2
46	= 44.
× 42	44^2 = 1,936
1,932	Subtract 1,936 − 4 = **1,932, the answer.**

To see it in a short form, think: $\frac{88}{2}$ = 44; 44^2 = 1,936; − 4 = **1,932.** Remember, this method works *only* on numbers with a difference of 4.

Multiplying Two Numbers Whose Difference Is 6

This time you square the average of the two numbers and subtract 9. Let's try 26 × 32:

	THINK: 26 + 32 = 58; 58 divided by 2
26	= 29.
× 32	29^2 = 841
832	Subtract 841 − 9 = **832, the answer.**

Once again, to see this in brief, think: $\frac{58}{2}$ = 29; 29^2 = 841; 841 − 9 = **832.**

Now You Can Do It

Remember to use the strategies for squaring as you find the short answers to these challenges:

44	18	26	65	19
× 38	× 12	× 22	× 61	× 16

84	28	76	15	52
× 81	× 26	× 74	× 14	× 51

7 A Hatful of Multiplication Tricks

SO FAR, WE'VE LEARNED several basic strategies for multiplying and squaring numbers. You can use those on any numbers that fit the particular method, but you also can add a bunch of tricks to your magic act, shortcuts guaranteed to make you look like a mathematical wizard.

MULTIPLYING BY 10, 100, ETC.

This one is so obvious it hardly seems worth mentioning, but knowing it makes multiplying and dividing decimals and fractions so easy, we'd better take note of it now.

Whenever you multiply a number by 10, you just add a 0 to it to get the answer. To multiply by 100, add two zeros. To multiply by 1,000, add three zeros. However many zeros are in the multiplier, you just add that to the number you're multiplying to get the answer. Like this:

$$\begin{array}{r} 12 \\ \times\ 10 \\ \hline 120 \end{array} \quad \begin{array}{r} 148 \\ \times\ 10 \\ \hline 1{,}480 \end{array} \quad \begin{array}{r} 148 \\ \times\ 100 \\ \hline 14{,}800 \end{array} \quad \begin{array}{r} 1{,}000 \\ \times\ 12 \\ \hline 12{,}000 \end{array}$$

$$\begin{array}{r} 1{,}000 \\ \times\ 148 \\ \hline 148{,}000 \end{array} \quad \begin{array}{r} 148{,}978{,}112 \\ \times\ 100{,}000 \\ \hline 14{,}897{,}811{,}200{,}000 \end{array}$$

ASTRONOMICAL FYI

As you can see, it's easy to get into astronomical figures playing with these numbers. Sometimes when you're reading about such things as how far apart the stars are, you'll see a number written like this:

$$18 \times 10^{12}$$

It means "18 with 12 zeros after it." That is, 18 followed by 12 zeros. I'm not even going to *try* to write that out!

For the fun of it, though, we could write any number followed by a zero using the same method. If we try it on the answers to the examples we used above, we get:

12×10^1 148×10^1 148×10^2 12×10^3 148×10^3
$148{,}978{,}122 \times 10^5$

The star nearest to earth is 4.3 light-years away from us. A light-year is the distance that light travels in one year, zipping along at 186,000 miles a second—about 5,878,000,000,000 miles. So that star is 5,878,000,000,000 × 4.3 miles from earth, or $25{,}275.4 \times 10^6$ miles.

Twenty-five trillion miles. That's spacey stuff! Math is like a spaceship—it can take you right into the cosmos.

MULTIPLES OF TENS, HUNDREDS, THOUSANDS . . .

Coming back down to earth, you can apply the same strategy to multiples of 10, 100, and the like. Make the number that ends in 0 the multiplier and the other number the multiplicand. Multiply the multiplicand by the first part of the multiplier. That gives you a base number that is the first part of the answer. The last part of the answer is the number of zeros in your multiplier. Like so:

12	**THINK:** 12 × 2 = 24.
× 20	Tack on a zero to get **the answer: 240.**
240	

12	**THINK:** 12 × 2 = 24.
× 200	Add two zeros for **the answer: 2,400.**
2,400	

Our base number in 12 × 20, which is 24, is really 24(0). All we have to do is remove the parentheses to get the answer. When we multiplied 12 × 200, our base number of 24 was really 24(00). And yes, the answer proved to be 2,400.

12	**THINK:** 12 × 8 = 96.
× 80	That's really 96(0), so the answer is **960.**
96(0)	

440	**THINK:** 12 × 44 = 528.
× 12	That's really 528(0), or **5,280.**
528(0)	

3,300	12 × 33 = 396.
× 12	That's really 396(00), or **39,600.**
396(00)	

Now You Can Do It

Here, too, the trick is to remember your place values. Try multiplying these figures:

241	3,672	998	4,600	96	4500
× 20	× 100	× 20	× 18	× 90	× 300

421 billion is the same as 421 $\times 10^9$. Express these figures using the same method:

9,876,000,000,000
18,000,000,000,000
47 trillion
13 quadrillion
52 googol (a googol is a 1 followed by 100 zeros)

MULTIPLYING BY 11

There are two good ways to multiply a number by 11. The strategy that I think is easiest goes like this: Take the number to be multiplied and add ten times the number.

426	**THINK:** 426	426
× 11	Plus 10 × 426	+ 4,260
		4,686

1,332	**THINK:** 1,332	1,332
× 11	Plus 10 × 1332	+ 13,320
		14,652

12	**THINK:** 12	12
× 11	Plus 10 × 12	+ 120
		132

You may have noticed something odd in the last example. The middle digit of the answer is the sum of the two digits in the number to be multiplied. The first and the last digits in the answer are the same as the first and last digits in the original number! That is, if you take the 1 and the 2 in 12 and add them (1 + 2 = 3), and then you stick that number in between the digits in 12, you get the answer: **132.**

You guessed it—that's *the second method for multiplying by 11.* There's a trick to it. You add the two digits and insert the sum between the digits. Take 18 × 11:

18	**THINK:** 1 + 8 = 9.
× 11	Put the sum between the two digits: 198.
198	Check it on your calculator: **198.**

Let's try it with 99 × 11.

99	**THINK:** 9 + 9 = 18.
× 11	Put the sum between the two digits: 9189.
oops!	Multiply 11 × 99 on your calculator: **1,089!**

Where did we go wrong?

When we added 9 + 9, we got a two-digit number. When that happens, you have to add that number's left digit to the left digit in the original number. That is, to get the right answer we needed to carry the 1 in 18 over to the first 9 in 99:

99	9 + 9 = 18.
× 11	Put the two-digit number in-between: 9 18 9.
	Then add the 1 to the left-hand 9: 9 + 1 = 10. That changes the first 9 to a 10 and gives
1,089	you **the correct answer: 1,089.**

Let's try it again on a smaller number:

37	3 + 7 = 10.
× 11	Place the 10 between the 3 and the 7: 3 10 7.
407	Then add the 1 to the 3: **407.**

To recap the two methods for multiplying by 11:

One way is to multiply the number by 10 and then add the number.

The other is to add the digits of a two-digit number and place the sum between them, carrying to the tens column if necessary.

MULTIPLYING BY 21, 31, 41, AND OTHER NUMBERS ENDING IN 1

The theory behind the first method for multiplying by 11 is that it's easier to multiply by 10 than by numbers ending in something other than zero. As a matter of fact, it's easy to multiply by any number ending in zero, as we've seen. What you're really doing is setting aside one piece of the 11 things to be multiplied, multiplying the rest of them, and then adding it back in at the end. You can do that with any number ending in zero.

Using the method we learned for multiplying by multiples of 10, multiply the first number by the first digit in the multiplier and tack on a zero. Then add the multiplier to that.

	THINK: 23 × 2 = 46. With the zero,
23	that's 460.
× 21	Add 23 + 460 for
483	**the answer.**

That's pretty easy. Let's try it on 42 × 31.

42	42 × 3 = 126. Put on a zero: 1260.
× 31	Add 42 + 1,260 to get
1,302	**the answer.**

Once again, how did we do that? Multiply by the next closest number ending in zero—20 if your multiplier is 21, 30 for 31, 40, for 41, and so on. Then add the number to be multiplied.

Use the quick-math trick for multiplying by 10s while you're trying the strategy for numbers ending in one:

$$\begin{array}{rrrrrr} 18 & 12 & 27 & 66 & 82 & 124 \\ \times\ 11 & \times\ 41 & \times\ 31 & \times\ 61 & \times\ 51 & \times\ 21 \\ \hline \end{array}$$

MULTIPLYING BY NUMBERS THAT END IN 5

One thing you know when multiplying a number ending in 5: The answer will always end in 5 or 0!

That has nothing to do with the strategy, but it can be used for a quick check of your answers. The plan behind the strategy goes like this:

If we can change the number ending in 5 to one that ends in 0, it will be easy to multiply. Two times any number ending in 5 is always a number ending in 0. So the first step is to double the number that ends in 5.

Now that we've changed the multiplier, we can change the number to be multiplied (you will remember that it's called a *multiplicand*) so that the end result is the same as multiplying two unchanged numbers. We doubled the multiplier. Now we halve the multiplicand. When we double one number and halve the other, we get the same product that we would with the original numbers.

Let's experiment on some small numbers:

$$\begin{array}{rcr} 4 & \tfrac{1}{2} \text{ of } 4 = 2 & 2 \\ \underline{\times\ 5} & 5 \times 2 = 10 & \underline{\times\ 10} \\ \mathbf{20} & & \mathbf{20} \end{array}$$

Those of you who haven't studied fractions yet need only remember that to get half of a number, you divide the number by 2.

$$\begin{array}{rcr} 8 & \tfrac{1}{2} \text{ of } 8 = 4 & 4 \\ \underline{\times\ 5} & 5 \times 2 = 10 & \underline{\times\ 10} \\ \mathbf{40} & & \mathbf{40} \end{array}$$

Now, that seems to be working. What happens if we apply it to some larger numbers, say 42 × 15?

$$\begin{array}{r} 42 \\ \underline{\times\ 15} \\ \mathbf{630} \end{array} \quad \begin{array}{l} \tfrac{1}{2} \text{ of } 42 = 21 \\ 15 \times 2 = 30 \\ \end{array} \quad \begin{array}{r} 21 \\ \underline{\times\ 30} \\ \mathbf{630} \end{array}$$

This looks good! Now we can use it on larger or clumsier numbers without testing the answer the old-fashioned way.

$$\begin{array}{r} 84 \\ \underline{\times\ 35} \\ \end{array} \quad \begin{array}{l} \tfrac{1}{2} \text{ of } 84\text{:} \\ 2 \times 35\text{:} \\ \end{array} \quad \begin{array}{r} 42 \\ \underline{\times\ 70} \\ \mathbf{2{,}940} \end{array}$$

$$\begin{array}{r} 128 \\ \underline{\times\ 85} \\ \end{array} \quad \begin{array}{l} \tfrac{1}{2} \text{ of } 128\text{:} \\ 2 \times 85\text{:} \\ \end{array} \quad \begin{array}{r} 64 \\ \underline{\times\ 170} \\ \mathbf{10{,}880} \end{array}$$

If the multiplicand is an odd number, the strategy still works, but you have to multiply with a fraction. Beginning human calculators may need to learn about fractions and decimals before taking this on.

Suppose you want to multiply 19 × 25:

$$\begin{array}{r} 19 \\ \underline{\times\ 25} \\ \end{array} \quad \begin{array}{l} \tfrac{1}{2} \text{ of } 19\text{:} \\ 2 \times 25\text{:} \\ \end{array} \quad \begin{array}{r} 9\tfrac{1}{2} \\ \underline{\times\ 50} \\ \mathbf{475} \end{array}$$

For those of you who haven't learned fractions yet, an easy way to solve this problem is by taking half of 50 (that's 25) and adding it to 9 × 50 (that's 450; so, 450 + 25 = **475**). To make a long story short, when you multiply a number by ½, you actually divide it by 2 (½ × 50 = 50 ÷ 2 = 25).

Let's try that again:

67	½ of 67: 33½
× 35	2 × 35 = 70. THINK: 33 × 70 = 2,310, plus ½ of 70, which is 35, is
2,345	2,310 + 35 = **2,345.**

Now You Can Do It

This strategy takes some practice, partly because you're using fractions for the first time, and partly because you sometimes have to perform three math processes at once: adding, multiplying, and dividing. Once you have it under control, though, you can work much faster than you can when you write down and add up offset rows of figures.

First, experiment with multiplying by ½:

4 × ½ =

8 × ½ =

26 × ½ =

54 × ½ =

108 × ½ =

Now continue the warmup with some mixed numbers:

4 × 2½ =

8 × 3½ =

26 × 4½ =

54 × 5½ =

108 × 6½ =

With these secrets mastered, the rest is easy.

$12 \times 5 =$

$12 \times 45 =$

$82 \times 35 =$

$36 \times 55 =$

$102 \times 15 =$

There are some special cases for multiplying numbers that end in 5. For example:

Multiplying a Number by 15

You can change a number that you want to multiply by 15 into a number that you can multiply by 10 to get the same answer. How? Just add half of the number to itself.

12	**THINK:** Half of 12 is 6; 12 + 6 =	18
× 15	Now multiply by	× 10
	The answer is:	**180**

It isn't magic; it's logic. Fifteen is half again as much as 10 (that is, it's 10 plus ½ of 10, or 5). So if you add half your multiplicand to the multiplicand and reduce your multiplier of 15 to 10 (taking half of 10 (or 5) *away* from the 15), you end up with the same answer.

When you perform an operation on one part of a math problem, the answer remains the same *if* you perform the *opposite* operation on another part of the problem. Here, we have added a half and subtracted a half.

Let's try this on some larger numbers:

44	Half of 44 is 22; 44 + 22 =	66
× 15	Multiply by 10:	× 10
	The answer:	**660**
116	Half of 116 is 58; 116 + 58 =	174
× 15	Multiply by 10:	× 10
	The answer:	**1,740**

Of course, with this strategy you're also apt to find yourself working with fractions. That's because half of an odd number is always a number and a half.

17	Half of 17 is 8½; 17 + 8½ =	25½
× 15	Multiply by 10:	× 10
	That's 25 × 10 + ½ of 10:	**255**

Here's another way to do this: Multiply the number by 10 and then add half the product.

44	THINK: 44 × 10 =	440
× 15	(Plus half of 440:)	+ 220
		660

This version makes working with odd numbers much simpler:

17	THINK: 17 × 10 =	170
× 15	(Plus half of 170:)	+ 85
		255

Multiplying a Number by 45

To do this, we notice that 5 is $\frac{1}{10}$ of 50. We're going to change the multiplier, 45, to 50. We'll multiply by 50, and then, to change back to 45, we'll subtract $\frac{1}{10}$ of the product from itself. What we're doing is multiplying by *50 − 5,* which is the same as multiplying by 45.

Let's try it out:

12	THINK: Multiply by 50. 12 × 50 =	600
× 45	$\frac{1}{10}$ of 600 = 600 divided by 10:	− 60
	The answer:	**540**

Multiplying a Number by 55

Just reverse this strategy when you want to multiply by 55. Now your multiplier is $\frac{1}{10}$ *more* than 50, instead of $\frac{1}{10}$ less. So you'll add $\frac{1}{10}$ of the product of 50 × the multiplicand, instead of subtracting it. This time, you're multiplying by *50 + 5.*

12	THINK: Multiply by 50. 12 × 50 =	600
× 55	$\frac{1}{10}$ of 600 = 600 divided by 10:	+ 60
	The answer:	**660**

Now You Can Do It

Remember: to get ½ of a number, divide by 2. To multiply a number by ½, do the same—divide by 2. To get 1/10 of a number, divide by 10. If fractions confuse you, work on the product instead of the multipliers whenever possible.

14	75	114	23	58
× 15	× 15	× 15	× 45	× 45

142	19	64	121
× 45	× 55	× 55	× 55

MULTIPLYING BY NUMBERS THAT END IN 9

You can apply the same strategy you used on 45 and 55. The idea is to fix things so you can multiply by 10, instead of the head-achey 9.

Well, 9 is one less than 10. So you can multiply a number by 10 and then subtract the number from the product to get the answer.

12	THINK: Multiply 12 × 10:	120
× 9	Now subtract 12:	− 12
	The answer:	**108**

Does it work on larger numbers? Sure!

56	56 × 10:	560
× 9	Minus 56:	−56
	The answer:	**504**
145	145 × 10:	1,450
× 9	Minus 145:	− 145
	The answer:	**1,305**

Multiplying by Two-Digit Numbers That End in 9

You use the same principle as in multiplying by numbers ending in 1: Change the multiplier so that it ends in 0, multiply by the digit in the multiplier's tens column. Now reverse the strategy and subtract the multiplicand from the result.

Multiplicand 12		12
Multiplier × 19	Add 1 to get 20:	× 20
		240
Subtract the 12:		− 12
The answer:		**228**

12		12
× 99	Add 1 to 99:	× 100
		1200
Subtract 12:		− 12
The answer:		**1,188**

45		45
× 59	Add 1 to 59 to get 60:	× 60
		2700
Subtract 45:		− 45
The answer:		**2,655**

Multiplying by Multiples of 9 (Up to 9 × 9)

Amazingly, every multiple of 9, up to 81, is the same as the next number higher that ends in 0 minus ⅒ of that number. Nine, for example, is ⅒ of 90, which is 9 more than 81 (which is 9 × 9).

This odd fact allows us to do the same thing with these numbers as we did when we wanted to multiply by 45.

First, let's review the multiples of 9:

9 × 2 = 18

9 × 3 = 27

9 × 4 = 36

9 × 5 = 45

9 × 6 = 54

9 × 7 = 63

9 × 8 = 72

9 × 9 = 81

If you recognize the numbers 18, 27, 36, 45, 54, 63, 72, and 81 like old friends, you can amaze your social circle with this trick.

Multiply 12 × 36.

12		12
× 36	Go to the next highest number in 0:	× 40
		480
Subtract ⅒ of the product:		− 48
The answer:		**432**

Any kid, taught this trick, could startle the fourth-grade teacher! To really amaze his teacher, he could extend the practice to multiplying by multiples of 99.

Multiplying by Multiples of 99

The same strategy works here, except that instead of subtracting $\frac{1}{10}$, you subtract $\frac{1}{100}$. Try it like this:

12		12
× 297	297 = 99 × 3; go to the next number in zero:	× 300
		3,600
Subtract $\frac{1}{100}$ of 3600, or		− 36
And **the answer** is:		**3,564**

The multiples of 99 are 198, 297, 396, 495, 594, 693, 792, 891, and 990.

Now You Can Do It

Can you see a pattern in the multiples of 9? How about in the multiples of 99? What is it?

See how many teachers or friends you can surprise by performing these tricks in your head:

$$\begin{array}{r} 19 \\ \times\ 36 \\ \hline \end{array} \qquad \begin{array}{r} 42 \\ \times\ 54 \\ \hline \end{array} \qquad \begin{array}{r} 27 \\ \times\ 18 \\ \hline \end{array} \qquad \begin{array}{r} 31 \\ \times\ 99 \\ \hline \end{array}$$

$$\begin{array}{r} 495 \\ \times\ 14 \\ \hline \end{array} \qquad \begin{array}{r} 297 \\ \times\ 23 \\ \hline \end{array} \qquad \begin{array}{r} 891 \\ \times\ 36 \\ \hline \end{array}$$

8 Subtraction

WE ARE TAUGHT IN school to think of subtraction as a negative process—a taking away of one thing from another. But to my mind, that's not what subtracting is all about.

It's really *addition in reverse*.

What I mean is this: When someone asks you for the answer to 74 minus 53, don't think, "What is 74 take away 53?" Instead, think, "What number, added to 53, equals 74?"

It's actually one of the first steps you'll learn in algebra. You're rephrasing the problem in the form of an algebraic formula: $53 + x = 74$. The letter x stands for the words "what number" in the question.

Here's how it works:

$$\begin{array}{r} 74 \\ -\ 53 \\ \hline \end{array}$$

Let's start with the tens column. We need to take 5(0) from 7(0)—that's 5 tens from 7 tens. Ask yourself, "If we added something to 5 tens to get 7 tens, what would it be?" Well, 5 tens plus 2 tens equals 7 tens.

$$\begin{array}{r} 7(0) \\ -\underline{5(0)} \\ \mathbf{2}(0) \end{array}$$

Add 2(0) to 5(0) to get 7(0):

2 is the first digit in the answer.

$$\begin{array}{r} 74 \\ -\underline{\ 53} \\ \mathbf{2}? \end{array}$$

Now move to the units column. What number added to 3 makes 4? It's 1.

$$\begin{array}{l} (0)4 \\ \underline{(0)3 \uparrow} \\ (0)\mathbf{1} \uparrow \end{array}$$

$$\begin{array}{r} 74 \\ -\underline{\ 53} \\ \mathbf{21} \end{array}$$

Write that in the units column and see **the answer: 21.**

Now You Can Do It

Ask yourself what number added to the bottom number makes the top number in these examples.

$$\begin{array}{r} 18 \\ -\underline{\ 9} \end{array} \quad \begin{array}{r} 26 \\ -\underline{14} \end{array} \quad \begin{array}{r} 199 \\ -\underline{150} \end{array} \quad \begin{array}{r} 54 \\ -\underline{14} \end{array} \quad \begin{array}{r} 75 \\ -\underline{25} \end{array} \quad \begin{array}{r} 263 \\ -\underline{161} \end{array}$$

That was easy. But what if you have to borrow from the tens to get the answer in the units column?

$$\begin{array}{r} 71 \\ -\underline{53} \\ \mathbf{2}(0) \end{array}$$

THINK: What number added to 5(0) equals 7(0)? It's 20. What number added to 3 equals 1? Uh oh. That doesn't work!

Luckily, we have a *key to human calculating:* We understand numbers. Seven tens and 1 unit is really the same as 6 tens and 11 units. That's because 11 is really 1 ten and 1 unit. Visualize that:

71 is actually		7(0)
	and	1
		7 1
6(0) and 11 is actually		6(0)
	and	1(0)
	and	1
		7 1

So, let's try changing 7(0) and 1 to 6(0) and 11, in our heads. Now the question is "What number added to 3 equals 11?" The answer is 8. But remember, when we changed 71 to 6(0) plus 11, we stole a 10 from 7(0). So we have to take a 10 away from the tens column in the answer, which started with a 2(0). Now it's a 1(0). So our answer combines 1(0) and 8, to make **18.**

I hear it now: "Scott! That's too complicated for my brain." Don't panic—you can make this really easy by looking and thinking before you start to figure. Here's the thought process:

$$\begin{array}{r} 71 \\ -53 \\ \hline \end{array}$$

THINK: I can't add anything to 3 to get one unit. I need some tens. Take one from the tens column.

$$\begin{array}{r} 71 \\ -53 \\ \hline \end{array}$$

THINK: this means 6(0) and 11
−5(0) and 3

What number added to 5(0) makes 6(0)? **1**(0)
What number added to 3 makes 11? **8**

$$\begin{array}{rr} 6(0) & \text{and } 11 \\ \uparrow \underline{-5(0)} & \underline{-\ 3}\uparrow \\ \uparrow\ 1(0) & 8\uparrow \end{array}$$

Now think: one 10 and 8 units are **18.**

$$\begin{array}{r} 71 \\ -53 \\ \hline \mathbf{18} \end{array}$$

The answer

Now You Can Do It

Remember that you can change tens to units by stretching your imagination to visualize two-digit numbers in the units column. When you borrow a ten from the tens column and give it to the units, you end up with more than 9 units in the right-hand column. Try your talents on these:

$$\begin{array}{r} 14 \\ -\ 8 \\ \hline \end{array} \qquad \begin{array}{r} 22 \\ -16 \\ \hline \end{array} \qquad \begin{array}{r} 60 \\ -44 \\ \hline \end{array} \qquad \begin{array}{r} 33 \\ -15 \\ \hline \end{array} \qquad \begin{array}{r} 96 \\ -78 \\ \hline \end{array} \qquad \begin{array}{r} 47 \\ -19 \\ \hline \end{array}$$

Let's try this on some larger numbers. First, an easy one:

898 −642	Glance at the bottom numbers and see that they all can add up to the numbers above them.
2 (00)	**THINK:** 6(00) and what make 8(00)? It's **2**(00).
25(0)	4(0) and what make 9(0)? That's **5**(0).
256	2 and what make 8? **6.** The **answer** is **256.**

Now for a challenger:

242
−168

6 + ? = 4 8 + ? = 2	Glance at the bottom numbers. You see that 1(00) and another number will add up logically to 2(00), but you can't add any number to 6(0) and come up with 4(0), nor can you add any number to 8 and get 2. You'll have to shuffle hundreds *and* tens.

24(0) = 1(00) + 14(0)

THINK: 2(00) is the same as 1(00) and 10 tens. Add the 10 tens to 4 tens for 14(0), which leaves 1(00) in the hundreds column. Okay. I can add something to 6 to get 14. But I still have a problem in the units column.

14(0) = 13(0) + 10 ones

THINK: 14(0) is the same as 13(0) and 10 ones. Ten ones and 2 units is the same as 12. Yes—I can add something to 8 to get 12. I'm going to work with 2 hundreds, 13 tens, and 12 ones. Visualize:

242 = 1(00) and 13(0) and 12
− 168 = 1(00) ↑ 6(0) ↑ 8 ↑

0(00)	**THINK:** What plus 1(00) equals 1(00)? Must be **0**(00).
7(0)	What plus 6(0) makes 13(0?) **7**(0)
74	What plus 8 equals 12? **4.** Put the two together for **the answer: 74.**

If you prefer, visualize the way you get the answer like this:

242	= 1(00) and	13(0)	and 12
−168	= 1(00) and	6(0) ↑	and 8 ↑
		7(0) ↑	4 ↑

Combine the columns for the answer: **74.**

Now You Can Do It

Remember the fifth key to human calculating: creativity. Subtraction is a very creative activity, in which you imagine yourself borrowing numbers and invent numbers that are smaller than zero. Picturing the process in your mind helps a lot. Now picture these:

456	283	692	505	927	1,063
−127	−168	−486	−333	−788	−984

THINKING ABOUT SUBTRACTION

How is subtraction different from addition? Even though I've called it "reverse addition," there's one thing you can't do with subtraction that you can with addition. It's this: You can't subtract in just any order. For example, 6 + 9 is the same as 9 + 6. But 9 − 6 is *not* the same as 6 − 9.

Sure, you can take a larger number away from a smaller one. The result is a *negative number*—it represents numbers less than zero. For example, when it's really cold in North Dakota, the temperature may be 40 degrees below zero on the Fahrenheit thermometer. That's a negative number. We write a negative number with a minus sign in front of it (−40), and we may read it as "minus 40."

To subtract a larger number from a smaller one, subtract the smaller number from the larger one, and then mark the answer as a negative number. To subtract $6 - 9$, first do $9 - 6 = 3$; then reverse it to $6 - 9 = -3$.

By the way, in a subtraction problem, the number that you are subtracting *from* is called the *minuend*. The number you are subtracting is the *subtrahend*, and the answer is the *remainder*.

$$\begin{array}{rl} 48 & \leftarrow \text{minuend} \\ \underline{-\ 2} & \leftarrow \text{subtrahend} \\ 46 & \leftarrow \text{remainder} \end{array}$$

Now You Can Do It

Let's see what actually happens when you try to reverse the direction of subtracting.

$$\begin{array}{cccccccc} 9 & 6 & 15 & 7 & 50 & 25 & 128 & 14 \\ \underline{-6} & \underline{-9} & \underline{-7} & \underline{-15} & \underline{-25} & \underline{-50} & \underline{-14} & \underline{-128} \\ 3 & -3 & & & & & & \end{array}$$

In some parts of the world, temperatures can drop 40 or 50 degrees in less than an hour. It was 35 degrees one afternoon in Fargo, North Dakota. By nightfall the mercury had fallen 47 degrees. So how cold *was* it that evening?

SUBTRACTING BY ADDING NUMBERS AROUND 100

When a number to be subtracted is close to 100, you can simplify the problem in the same way we simplified multiplication—by changing it to a number that ends in zero, and then converting the answer to get rid of the changes we made.

Watch this:

137	**THINK:** What is the difference between
− 96	100 and 96? 4. Remember that number.
137	Subtract 100 instead of 96.
−100	**THINK:** 100 plus what is 137? It's 37.
37	**THINK:** 96 is 4 less than 100; so,
+ 4	37 is 4 less than the real answer.
41	Add 4 to 37 to get **the answer: 41.**

SUBTRACTING BY ADDING TO EACH NUMBER

Here's a little secret that makes subtracting numbers simple: If you add an amount to both numbers in a subtraction problem, *the difference between the numbers stays the same.*

Let's try that on some small numbers.

Add 2 to both numbers in a problem:

$$\begin{array}{r} 5 + 2 = \quad 7 \\ \underline{-\ 2} + 2 = \underline{-\ 4} \\ \mathbf{3} \qquad\quad \mathbf{3} \end{array}$$

If we can do that, we can change a lot of numbers into figures that end in zero, and that will make our subtracting ever so much easier. Say:

7 + 3 =	10	7 is closer to a number ending in 0 than 3. If 3 is added to 7, we get 10. So,
− 4 + 3 =	− 7	add 3 to both numbers.
3	**3**	The answer is the same.
7 − 4 =	**3**	7 − 4 is the same as (7 + 3) minus
10 − 7 =	**3**	(4 + 3), or 10 − 7.

Let's try that on a couple of two-digit numbers. Remember that we will add the same amount to both numbers in our problem to make new numbers that are easier to work with. The answer will be the same. This time we'll try 47 − 34.

47 + 3 =	50	Of 47 and 34, 47 is closest to a number ending in 0—in this case, 50. We add 3 to 47 to get 50, and then add 3 to 34, which gives 37. The new, easy problem is 50 − 37, whose answer we can see at a glance: **13.**
−34 + 3 =	− 37 ↑	
13	**13** ↑	

Now for some serious calculating:

148 + 3 =	151	97 is 3 away from 100.
−97 + 3 =	− 100 ↑	Add 3 to both numbers.
	51 ↑	Subtract the easy numbers.

Test it on your calculator, and you will find that 148 − 97 is also 51.

Will it work on four-digit numbers? Sure.

4,873 + 8 =	4,881	Add 8, because . . .
− 3,492 + 8 =	3,500	8 added to 3,492 is 3,500.

Now let's look at this new problem:

4,881 − 3,500 **1**(000)	Start with the thousands columns: 4 (000) − 3(000) = **1**(000), giving us the first number in our answer. Visualize: 4(000) ↑ − 3(000) ↑ **1**(000)
8 81 5 00 **3**(00)	Now move to the hundreds column: 8(00) − 5(00) = 3(00), the second number in the answer. Visualize: 8(00) ↑ − 5(00) ↑ **3**(00)
1(0 00) + **3**(00) **1 3**(00)	Combine those two to visualize the first two numbers in the answer: **1**(000) + **3**(00) = **13**(00).
81 − 00 **81**	Next, subtract the tens and units columns. The subtrahend contains no values in the tens and units columns; those places are held by 0s. Subtract 0(0) from 81 to get 81, and 0 from 1 to get 1. Or, to put it more plainly, 81 minus nothing is 81. These are the last two digits in the answer.
13(00) + 81 **13 81**	All that remains is to combine the last two digits with the first two: **13**(00) + **81** = **the answer: 1,381.**

Incredibly easy. Check it with a calculator: 4,873 − 3,492 = 1,381.

Now You Can Do It

Try the two strategies above on these numbers near 100s or 1,000s:

186	214	540	358	927	7,642
− 98	− 189	− 396	− 292	− 583	− 4,397

SUBTRACTING BY SUBTRACTING FROM EACH NUMBER

Of course, if you can simplify a problem by adding to each figure to make one of them end in zero, you can do the same by subtracting to bring one of the figures down to a number ending in zero. For example:

1,486 − 2 =	1,484	Subtract 2, because . . .
− 12 − 2 =	− 10 ↑	12 is 2 more than 10.
	1,474 ↑	

Can that possibly be right? Punch it in to a calculator: 1,486 − 12 = 1,474.

It works with numbers near 100, too.

656 − 24 =	632
− 124 − 24 =	− 100 ↑
	532 ↑

By changing a figure that you are going to subtract into any number that ends in zero, you make your problem 100 percent easier.

$$
\begin{array}{rcl}
8{,}796 - 342 & = & 8{,}454 \\
\underline{-\ 6{,}342} - 342 & = & \underline{-\ 6{,}000} \uparrow \\
 & & \mathbf{2{,}454} \uparrow
\end{array}
$$

These strategies are so simple, you can look your friends straight in the face and come up with the answer before they can get it on their machine. All it takes is a little practice.

$$
\begin{array}{rcl}
98{,}387 - 246 & = & 98{,}141 \\
-\ 76{,}246 - 246 & = & \underline{76{,}000} \uparrow \\
 & & \mathbf{22{,}141} \uparrow
\end{array}
$$

$$
\begin{array}{rcl}
42{,}761 + 8 & = & 42{,}769 \\
-\ 18{,}992 + 8 & = & \underline{19{,}000} \uparrow \\
18{,}992 + 8 & = & \mathbf{22{,}141}
\end{array}
$$

Now You Can Do It

Wait till Mom sees you toss these things off as though they were as easy as . . . as easy as they are!

56	77	129	245	684
− 21	− 38	− 78	− 120	− 434

9 Division

IF SUBTRACTION IS REVERSE ADDITION, then division can be called "reverse multiplication." The trick is to find out how many times one number has to be multiplied to get to another number.

For example, $6 \div 3 = 2$ tells us the same thing as $2 \times 3 = 6$. It just says it in a different way. And $100 \div 50 = 2$, which is another way of saying $50 \times 2 = 100$. This fact makes it easy to check your answers in division—just multiply the answer (called the *quotient*) by the number you're using to divide (the *divisor*), and if the result is the same as the number you were dividing into (the *dividend*), the answer is correct.

People write about division in several ways. One is by using this symbol:

$$\overline{)}$$

$$\begin{array}{r} 2 \leftarrow \textbf{quotient} \\ \textbf{divisor} \rightarrow 50\overline{)100} \leftarrow \textbf{dividend} \end{array}$$

This statement reads "100 divided by 50 is 2." Or you could read it as "50 divided into 100 is 2." Or simply, "50 into 100 is 2."

You could write the same thing like this:

$$100 \div 50 = 2$$

And that reads "100 divided by 50 equals 2."

Division problems can also be written as fractions, like this:

$$\frac{100}{50} = 2$$

Again, you could say "100 divided by 50 equals 2." Sometimes people read this as "100 over 50 equals 2."

For this chapter, I am going to assume you know your multiplication and division tables through 12. You can help yourself learn these tables using ordinary index cards. Write the fact you want to learn on one side and the answer on the other.

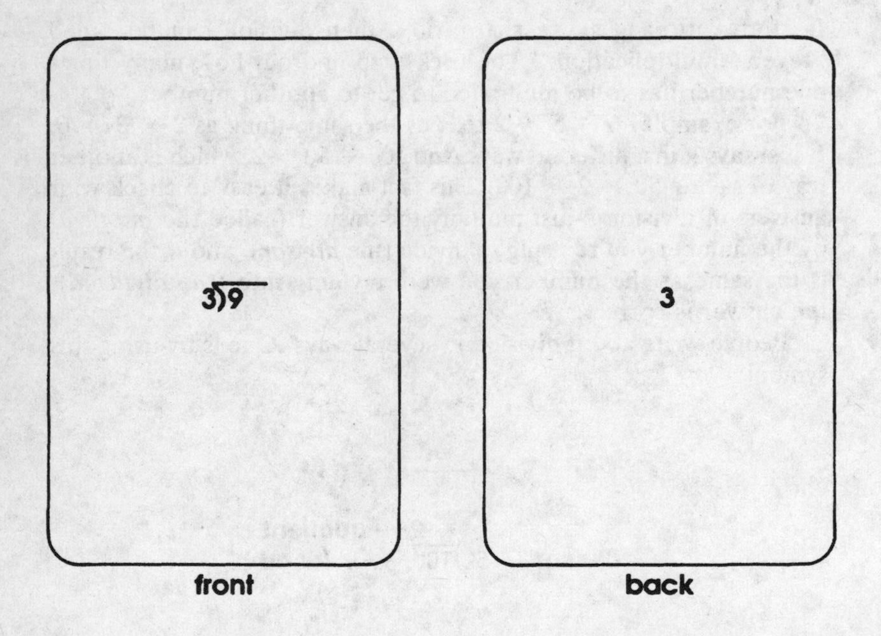

Once you know the multiplication and division tables, there is one easy way to simplify difficult-looking problems. It's based on the fact that you can get the same answer by dividing a number by *the parts* of a divisor as you do with the whole divisor. For example, you know that 18 is the same as 6 × 3, or as 9 × 2. These parts of a larger number are called *factors*. That is, 6 and 3 are factors of 18, and so are 9 and 2. To break a larger number up into its smaller parts is called *to factor* the number.

We can use these factors to our advantage.

Suppose we want to divide 486 by 18:

$$18\overline{)486}$$

It's easier to divide 486 by 6 and then again by 3 than to divide it by 18. First divide by one factor, then by the next, in whichever order is easiest. Let's try it:

$$6\overline{)486}$$

THINK: 6 does not divide into 4. But 6 × 8 is 48, so 48 ÷ 6 = 8. Write the 8 over the tens column in the dividend. Now, multiply 6 × 8. The answer is 48. Write that beneath the two digits you divided 6 into—the first two digits in the dividend (48). Subtract this answer from those two digits: 48 − 48 = a remainder of 0.

$$\begin{array}{r} 8 \\ 6\overline{)486} \\ \underline{48} \\ 0 \end{array}$$

Now bring down the 6 in the dividend and write it next to the 0. Divide 6 into that number: 6 × what = 6? 1. Write that in the ones column of the answer.

$$\begin{array}{r} 81 \\ 6\overline{)486} \\ \underline{48} \\ 06 \end{array}$$

Now we're half done: we still have to divide this result by the other factor of 18, which was 3:

$3\overline{)81}$

THINK: 3 goes into 8 two times, with a remainder of 2. Write the first 2 over the 8. Multiply 2 × 3 to get 6; write that under the 8 and subtract to see the remainder, 2. Bring the 1 in 81 down next to the remainder 2 to form a new number, 21. Divide 3 into 21 to get 7. Write it in the quotient over the ones column. Multiply 3 × 7 to get 21; write it beneath the first 21. Subtract for a remainder of 0.

$$\begin{array}{r} \mathbf{27} \\ 3\overline{)81} \\ \underline{6} \\ 21 \\ \underline{21} \\ 0 \end{array}$$

$$\begin{array}{r} \mathbf{27} \\ 18\overline{)486} \end{array}$$

This is **the answer: 27.**

$$\begin{array}{r} \mathbf{27} \\ 18\overline{)486} \end{array} = 486 \div 6 \div 3, \text{ because } 18 = 6 \times 3.$$

Before we go on to the next strategy, let's discuss what we did in the second part of the example above.

When you divide one number into another and it doesn't fit evenly—that is, there are numbers left over—the leftovers are called the *remainder.* For example, 5 does not divide into 18 evenly. The best we can do is 5 × 3 = 15, which is three short of 18. That is, 18 ÷ 5 = *3, with a remainder of 3.* The same thing happened when we tried to divide 8 by 3: the closest we could come was 3 × 2 = 6, which was two less than 8. So, 8 ÷ 3 = 2, with a remainder of 2.

Can you picture these numbers as concrete objects? Imagine the number 18 as a glass jar holding 18 pretzels. When you ask yourself how many times 18 can be divided by 5, you're really asking how many stacks you'd get if you took out the pretzels and piled them in stacks of 5. The answer, of course, is 3 stacks, with a smaller stack of 3 left over. In division, the fifth key to human calculating—*creativity*—can really help you. Visualizing your numbers is the best way to understand what goes on when you divide one figure by another.

When you get a remainder in the middle of a division problem—as we did in the 81 ÷ 3 step above (we divided 8 by 3 to get 2, with a remainder of 2)—it's called a *partial remainder.* The example above shows how you work with partial remainders—subtract from the part of the dividend you're working with and bring down the next number to the right; then divide again into that. Writing all the partial remainders down in this way is called *long division.*

Ordinarily, long division is used to help you keep track of problems that involve larger numbers. When you're dividing by numbers up to 12, you use *short division.* In that case, you either keep the partial remainder in your head or you make a note of it, like this:

2 3)81 **2**	**THINK:** 3 × what = 8? The closest is 3 × 2 = 6. Write the 2 under the tens column of 81 and, in your head, subtract the 6 from the 8 in 81, for a remainder of 2. Write this 2 over the 1 in 81.
8(0) − 6(0) = 2(0)	Remember: The remainder is in the tens column. It is *really* 2(0)!
2(0) + 1 = 21	Now, combine the partial remainder of 2(0) with the 1 in the ones column of 81 to get the new number to divide: 21.
2 3)81 **27**	Divide 3 into 21 to get 7. Because 7 × 3 = 21, and 21 − 21 = 0, there is no remainder. The answer is **27.**

Another way to look at a little problem like this is to ask yourself, "3 times what makes 80?" Well, you know that 3 times 30 will be 90, and 3 times 20 is 60, so the answer is somewhere between 20 and 30. Your base number can't be 30, so it must be 20. Put a 2 in the tens column of your answer and tack the digit in the ones column onto it: 21. Now divide 21 by 3 to get 7.

Let's try that with 87 ÷ 3:

2
3)87
29

THINK: 3 × what makes 80? Something larger than 20 and smaller than 30; the base number is 20. The first digit of the answer is 2. Tack on the 7 from the dividend's units column and divide again by 3: 27 ÷ 3 = 9. Put the 9 in the units column of the answer, and read **the whole answer: 29.**

Let's try the first method of short division on a larger number:

6)197,208

To keep things simple, let's start with the two left numbers in the dividend, which are in the one hundred thousand and the ten thousand columns. **THINK:** 6 × what makes 19? Something larger than 3 and smaller than 4. So we know our base number is between 30,000 and 40,000. Write a 3 under the ten thousands column and make a note of the partial remainder, which is 1, over the thousands column. Now, 6 × what makes 17(000)?

1
6)197,208
3

Well, 2 × 6 is 12 and 3 × 6 is 18; so we must want 2 with a remainder of 5; write the 2 under the thousands column and note the 5 over the hundreds column. Okay, 6 × what makes 52(00)?

1 5
6)197,208
32

It's something between 8 and 9; 6 × 8 is 48, giving us a result of 8 with a partial remainder of 4. Write the 8 under the hundreds column and note the 4 over the tens.

1 54
6)197,208
32 8

1 544 6)197,208 32 86	Next act: 6 × what = 4(00)? Something between 6 and 7; 6 × 6 = 36, leaving a partial remainder of 4, so we write the six under the tens column and note the 4 over the units.
1 544 6)197,208 **32,868**	And the last step: 6 × what = 48? A nice, even 8. Write it in the units column and read **the answer: 32,868.**

Now You Can Do It

Decide for yourself whether to use long or short division on these examples:

6)124   3)14,282   12)86   12)86,495   43)13,247

Now that we've reviewed the basic procedure for dividing numbers, let's look at a few high-speed strategies.

DECIMALS AND FRACTIONS

To use the first method, we first need to know what is meant by a *decimal point.*

A decimal point is a mark that looks like a period, which you sometimes see written in a number. It is a way of expressing a fraction. This fraction—that is, the part of a number—is a part of 10, a part of 100, a part of 1,000, and so on to infinity. The numbers that appear to the right of the decimal point are really fractions. For example:

.1 is really one-tenth (1/10).
.01 is one one-hundredth (1/100).
.001 is one one-thousandth (1/1,000).

.0001 is one ten-thousandth (1/10,000).
.00001 is one one-hundred-thousandth (1/100,000).

When we write dollars and cents as figures, we're actually using decimals. There are 100 pennies in a dollar—that's why we call a penny a "cent," from the Latin word for 100. So, one penny is one one-hundredth (1/100) of a dollar—and that's why we write it as $.01. Fifty pennies make 50/100 of a dollar, or $.50.

Percentage works exactly the same way. The term "one percent" means "one part of 100." That's the same as .01, or 1%, or 1/100.

This way of writing parts of 100 makes for some handy tricks. It gives us a standard way to express common fractions, and once we know what part of 100 these fractions are, we have an easy way to divide. Think about this:

1/4 of anything is the same as .25.
1/2 is the same as .50, or .5.
3/4 = .75
1/3 = .33 (approximately)
2/3 = .66 (approximately
1/8 = .125

These fractions can also be written as percentages, as decimal points, or as fractions of 100:

1/4 = 25% = .25 = 25/100
1/2 = 50% = .50 = 50/100 (This is the same as:
50% = .5 = 5/10.)
3/4 = 75% = .75 = 75/100
1/3 = 33% = .33 = 33/100
2/3 = 66% = .66 = 66/100
1/8 = 12.5% = .125 = 12.5/100

Using the decimal point, then, is just another way of talking about parts of 10, 100, 1,000, or more.

The remainder in a division problem can be expressed as a fraction or as a decimal. Instead of saying 8 ÷ 3 = 2 with a remainder of 2, we can express that remainder of 2 as a fraction. How? By putting the remainder (2) on the top of the fraction and the divisor (3) on the bottom, to give us ⅔. So, we could say, 8 ÷ 3 = 2⅔. The fraction ⅔ is approximately the same as the decimal .66—and so we could also say 8 ÷ 3 = 2.66.

Now, if you are beginning to understand this, you have a key to the next strategy:

DIVIDING BY 10, 100, 1,000, AND SO FORTH

Any time you divide a number by 10, you simply move the decimal point one place to the left.

What if the number doesn't have a decimal point? Just assume the decimal point is at the very end of the number. That is, the number 245 is the same as 245.(0).

To divide 245 by 10, move the decimal point from 245.(0) one place to the left, to 24.5.

245 ÷ 10 = **24.5,** or **24½.**

Let's test that with some short division:

```
     4
10)245
    2
```

THINK: 10 × what makes 24? 10 × 2 is 20; write the 2 beneath the tens column. Subtract 20 from 24 for a partial remainder of 4; write the 4 over the units column and divide 10 into 45. 10 × what makes 45? It must be 4, because 10 × 4 = 40, giving us **24 with a remainder of 5.**

```
     4
10)245
   24, r. 5
```

Five over 10 is 5 ÷ 10, or .5, or ½. That checks with our shortcut answer, **24.5.**

To divide by 100, you move the decimal point over two places to the left. To divide by 1,000, move it over three places. To divide by 10,000, move it four places. In other words, move the decimal points as many places to the left as there are 0s in the divisor. If you run out of figures, insert 0s between the decimal point and the first figure of the dividend.

245 ÷ 100 = 2.45
245 ÷ 1,000 = .245
245 ÷ 10,000 = .0245
245 ÷ 100,000 = .00245

That last number is pretty small: You read it as *two hundred forty-five hundred-thousandths!*

If you wanted to divide 876,942,235 by 1,000, you'd just move the decimal point from the end of the number three places to the left: The answer would be **876,942.235.** It's even easier when the number already has a decimal point in it, because then you don't have to remember to visualize the decimal point at the end:

> You charged $89.64 on your credit card for 10 identical dog collars to put on the family's pet pig. How much did you spend on each collar?
>
> $89.64 ÷ 10 = $8.964, or, rounded off, $8.96 apiece.

We will talk more about fractions, decimals, and percentages in Chapters 13 and 14.

Now You Can Do It

What fraction of a dollar is 50 cents? Express it as two different fractions and with a decimal point. Can you do the same for a quarter? How about six bits? ("Two bits" means 25 cents.)

If $.50 is 50% of a dollar, what percentage of a dollar is:

$.12 $.25 $.36 $.05 $1.00

And now, meet these challenges!

$10\overline{)24{,}876}$ $100\overline{)486{,}827.32}$ $10{,}000\overline{)923{,}674{,}163.88}$

$10\overline{)1.248}$ $100\overline{)486}$ $100\overline{)98}$

Canceling Zeros

As long as we're playing with numbers that end in zero, you might as well know that when any divisor ends in one or more zeros, you can cancel them out before you divide. It's a lot easier, for example, to divide by 26 than by 2,600.

To cancel zeros in a divisor, move the decimal point in the dividend one place to the left for each zero you cancel.

```
2,600)47,164

     18.14
  26)471.64
     26
     211
     208
       36
       26
       104
       104
         0
```

THINK: Cancel the zeros in 2,600 by moving the decimal point two places to the left. Now perform the required long division.

This helps because, in effect, you divided both the divisor and the dividend by the same number—100. When you did this, you ended up handling smaller, less clumsy numbers.

MULTIPLYING OR DIVIDING BOTH PARTS OF A DIVISION PROBLEM BY THE SAME NUMBER

When you multiply or divide the divisor and the dividend by the same number—any number—you don't change the answer to the division problem! That's why canceling zeros (dividing both parts by 100) works.

Let's try that on something simple:

$$2\overline{)4}$$ quotient 2

Let's multiply both parts by 3: 3 × 2 = 6 and 3 × 4 = 12. So the new problem is

$$6\overline{)12}$$ quotient **2**

12 ÷ 6. Yes—**the answer** is the same: **2.**

There are lots of possibilities here: If the divisor ends in 5, you can double it and the dividend, and then you can divide by a number ending in zero.

$$25\overline{)212}$$

THINK: Double both parts. 25 × 2 = 50; 212 × 2 = 424. The new problem is 424 ÷ 50. It's easier to divide by 50 than by 25; 5 into 42 is 8 and some, and so 50 goes into 424 about 8 times. **The answer** is **8.48**.

```
     8.48
50)424.00
   400
    24 0
    20 0
     4 00
     4 00
        0
```

If the divisor is a multiple of 5, you can divide it and the dividend by a factor that will turn the divisor into a 5. This gives you a very easy division problem:

$$45\overline{)720}$$

THINK: 45 is the same as 9 × 5. Divide both parts by 9: 45 ÷ 9 = 5; 720 ÷ 9 = 80. The new problem is 80 ÷ 5. Short division shows us **the answer** is **16**.

```
  3
5)80
  16
```

We would have had to use long division to divide 720 by 45.

Now You Can Do It

$35\overline{)98}$ $15\overline{)860}$ $75\overline{)390}$ $16\overline{)48}$

10 A Hatful of Division Tricks

POTENTIAL HUMAN CALCULATORS LOVE division because it's full of possibilities. When you know how numbers work, you can find all sorts of easy shortcuts that take you straight to the answer to some pretty challenging-looking problems. While others are sweating over long division, you can arrive at the answer in the blink of an eye.

Try these strategies on your friends and teachers:

DIVIDING BY 5

To divide a number by 5, multiply it by 2 and then divide the result by 10. For example, divide 242 by 5:

$5\overline{)242}$ $\mathbf{48.4}$ $5\overline{)242}$	THINK: 242 × 2 = 484. To divide by 10, move the decimal point one place to the left: **48.4 is the answer.**

Too easy! Can we possibly believe this is true without knowing the pain of dividing the hard way?

4 5)242 **48, r. 2**	Five goes into 24 four times with a remainder of 4. Five into 42 is 8, with a remainder of 2. And 2 over 5 (2/5) is the same as .4, so 48 with a remainder of 2 equals **48.4.**

Dividing a Number by 50

To divide by 50, multiply the number by 2 and divide it by 100. The principle is the same as dividing by 5:

6.82 50)341.00	THINK: 341 × 2 = 682. To divide by 100, move the decimal point two places to the left: **The answer is 6.82.**

And of course this works with the next biggest step:

Dividing a Number by 500

This time, multiply by 2 and divide by 1,000.

2.482 500)1,241.000	THINK: 1,241 × 2 = 2,482. To divide by 1,000, move the decimal point three places to the left, for **the answer: 2.482.**

Dividing by 25

Multiply the number by 4 and divide by 100.

$$60 \div 25 = 60 \times 4 \div 100 = 240 \div 100 = \mathbf{2.4}$$

Dividing by 250

As in the previous example, multiply the number by 4 and divide by 1,000.

$$600 \div 250 = 600 \times 4 \div 100 = 2400 \div 1,000 = 2.4$$

Dividing by 125

This strategy is simple, as long as you know your 8s multiplication table. To divide a number by 125, all you have to do is multiply the number by 8 and divide by 1,000.

$$125\overline{)310} = 8 \times 310 \div 1,000 = 2,480 \div 1,000 = 2.480$$

Now You Can Do It

Try the strategies you've learned on these test cases:

$5\overline{)309}$ $50\overline{)1804}$ $500\overline{)3416}$ $25\overline{)360}$ $250\overline{)431}$ $125\overline{)4868}$

HOW TO FIGURE OUT WHETHER A NUMBER CAN BE DIVIDED EVENLY

One of the big headaches in dividing is trying to know whether a number can be divided evenly by another. There are several rules of thumb that will help you determine whether a number is evenly divisible, and if so, by what.

Odd numbers can *only* be divided evenly by another odd number. That is, if the number ends in 1, 3, 5, 7, or 9, you know it can only be divided evenly by a number that ends in 1, 3, 5, 7, or 9. It *can't* be divided evenly by a number ending in 0, 2, 4, 6, or 8.

Even numbers can be divided evenly by either odd or even numbers.

If you try to divide an odd number—such as 15—by an even number—such as 6—the answer will always have a remainder (a fraction).

Can It Divide by 2?

Numbers that end in 0, 2, 4, 6, or 8 are evenly divisible by 2.

Can It Divide by 3?

Add up the number's digits. If the sum is evenly divisible by 3, then the number can be divided by 3.

Can we divide 786 by 3 with no remainder?

$7 + 8 + 6 = 21$

$21 \div 3 = 7$

THINK: add the digits. Yes, 21 is evenly divisible by 3. Therefore, 786 is divisible by 3. Short division proves it.

```
    1
  3)786
    262
```

Can It Divide by 4?

Look at the last two digits. If they are both zeros, or if they form a two-digit number evenly divisible by 4, then 4 will go evenly into the whole number.

Can we divide 7,600 by 4 with no remainder?

7600

THINK: This number ends in two zeros. It should divide by 4. Prove it by short division.

```
   3
 4)7600
  1,900
```

Can we divide 1984 by 4 with no remainder?

1984

THINK: 84 ÷ 4 = 21. This will surely work. In short division, it does!

```
   32
4)1984
  496
```

Can It Divide by 5?

Any number that ends in 5 or 0 is evenly divisible by 5. This seems obvious to anyone who can count by 5s: 5, 10, 15, 20, etc.

Can we divide 1,874,295 by 5 with no remainder?

1,874,295

THINK: This number ends in 5. So, it must divide evenly by 5. Short division shows it does.

```
   32 424
5)1,874,295
  374,859
```

Can It Divide by 6?

See if the number is even. If it is not, it cannot be divided evenly by 6. If it is, add up the digits; if their sum is evenly divisible by 3, the number can be evenly divided by 6.

Can we divide 27,354 by 6 with no remainder?

27,354

2 + 7 + 3 + 5 + 4 = 21

THINK: Is this number even? Yes. The total of its digits is 21; 21 ÷ 3 = 7. The number 27,354 is evenly divisible by 6.

```
6)27,354
   4,559
```

Can It Divide by 7?

Figuring out whether a number is evenly divisible by 7 is so complicated that it is easiest to simply try dividing by 7. Sorry.

Can It Divide by 8?

Look at the last three digits. If they are all zeros, or if they form a number that is evenly divisible by 8, then the whole number can be divided evenly by 8.

Can we divide 19,000 by 8 with no remainder?

19,000

THINK: The last three digits are zeros. We can divide the number evenly by 8. Short division proves this guess to be correct.

```
   364
8)19,000
  2,375
```

Can we divide 14,192 by 8 with no remainder?

14,192

THINK: The last three digits form a number that is evenly divisible by 8: $192 \div 8 = 24$. This should work. With short division, we find an even answer.

```
   653
8)14,192
  1,774
```

Can It Divide by 9?

Add up the number's digits. If their sum is evenly divisible by 9, so is the number.

Can we divide 3,618 by 9 with no remainder?

3,618 3 + 6 + 1 + 8 = 18 $9\overline{)3,618}$ **402**	**THINK:** The sum of the digits is 18; 18 ÷ 9 = 2. So, 3,618 should be evenly divisible by 9. Short division shows this to be true.

Can It Divide by 10?

Any number that ends in 0 is evenly divisible by 10. Intuition tells you this when you count by 10s: 10, 20, 30, 40, etc. Every multiple of 10 ends in 0—division is reverse multiplication, so every number ending in 0 must divide by 10.

Can It Divide by 11?

If all the digits in a number are the same (such as 11 or 222), *and* it has an *even number of digits,* the number will divide by 11 with no remainder.

$$77 \div 11 = 7; \text{ but } 777 \div 11 = 70.63$$

Why? Because 77 has two digits, an *even number,* but 777 has three digits, an *odd number* of digits. It doesn't matter that 77 and 777 are odd numbers; what matters is the *number of digits* in the numbers.

$$4{,}444 \div 11 = 404; \text{ but } 44{,}444 \div 11 = 4040.36$$

Count the digits: 4,444 contains four digits, an even number. And 44,444 has five digits: an odd number. Four digits: divisible by 11; five digits: not evenly divisible by 11.

If the digits are different, start on the right and count the digits. Now, add the digits that are in the "odd" slots, and then add the digits in the "even" slots.

3 0 4 , 0 6 2
↑ ↑ ↑ ↑ ↑ ↑
6 5 4 3 2 1

THINK: The numbers in the "odd" positions are 2, 0, and 0; the numbers in the "even" position are 6, 4, and 3.

Now, add up the numbers in the "odd" places and the numbers in the "even" places.

$2 + 0 + 0 = 2$

$6 + 4 + 3 = 13$

The "odd" digits total 2. The "even" digits total 13.

Subtract the smaller number from the largest. If the difference is evenly divisible by 11, then so is the number you started with.

$13 - 2 = 11$

```
     8 742
11)304,062
    27,642
```

11 is evenly divisible by 11: 11 ÷ 11 = 1. Short division shows the answer is an even **27,642.**

To see whether a three-digit number can be evenly divided by 11, add the two outside digits. If their sum is the same as the middle digit, or if the differences between their sum and the middle digit is 11, then 11 will go evenly into the number.

572

2
11)572
52

THINK: 5 + 2 = 7. That's the same as the middle digit; so, 572 must be evenly divisible by 11.

924

9 + 4 = 13

13 − 2 = 11

4
11)924
84

THINK: 9 + 4 = 13. Subtract the middle digit from this sum to get 11; the number must be evenly divisible by 11. Try short division to prove it.

The strange fact about this trick is that if you reverse the outside digits, the new number will also be evenly divisible by 11. Using our previous examples of 572 and 924:

5
11)275
25

9
11)429
39

Can It Divide by 12?

If a number can be evenly divided by 3 *and* by 4, it can also be evenly divided by 12. Use your strategies for evenly dividing by 3 and 4 to determine this:

508,272

5 + 0 + 8 + 2
+ 7 + 2 = 24

24 ÷ 3 = 8

72 ÷ 4 = 18

2 467
12)508,272
42,356

THINK: The sum of the digits is 24; 24 ÷ 3 = 8; the number is evenly divisible by 3. The last two digits, 72, are evenly divisible by 3. This number must be evenly divisible by 12.

Can It Divide by 15?

If a number can be evenly divided by 3 *and* by 5, it can also be evenly divided by 15. Here again, use your strategies for dividing by 3 and by 5 to decide whether you can divide evenly by 15:

390

$3 + 9 = 12$

$390 \div 5 = 78$

THINK: The sum of the digits is 12, a number that is evenly divisible by 3. The number ends in 0; so, it is evenly divisible by 5. Therefore, it must also be evenly divisible by 15.

```
   29
15)390
   26
```

SEVEN MORE DIVISIBILITY STRATEGIES

You can determine whether numbers are evenly divisible by larger numbers, too. Try these tricks:

A number can be evenly divided *by 20* if the units digit is 0 and the tens digit is even.

A number can be evenly divided *by 22* if it is even and also evenly divisible by 11.

A number can be evenly divided *by 24* if you can divide it evenly by 3 *and* by 8.

A number that ends in 00, 25, 50, or 75 can be evenly divided *by 25*.

A number can be evenly divided *by 30* if it ends in 0 and is evenly divisible by 3.

A number can be evenly divided *by 33* if you can divide it evenly by 3 *and* by 11.

A number can be evenly divided *by 36* if you can divide it evenly by 4 *and* by 9.

Now You Can Do It

Which of these numbers can be divided evenly by 3?

13 126 1,704 145 168 171

Which of these numbers can be divided evenly by 4?

96 154 178 236 944 3,780

Which of these numbers can be divided evenly by 5?

15,670 25,465,256 42,345 95,253 12,000

Which of these numbers can be divided evenly by 6?

270 3,954 2,956 3,996 25,452 39,359

Which of these numbers can be divided evenly by 8?

576 2958 31,536 4,676 23,654,123

Which of these numbers can be divided evenly by 9?

369 693 963 431 8,847 59,339

Which of these numbers can be divided evenly by 11?

333 714 9,856 10,858 7,194 3,533

Which of these numbers can be divided evenly by 12?

488 9,468 7,842 5,472 11,556 3,092

Which of these numbers can be divided evenly by 15?

11,115 12,943 2,310 24,125 945 147,795

Figure out how to divide each of the following numbers evenly (numbers divisible by 10, 5, or 2 are also evenly divisible by larger figures):

23,775
504
52,272
64,614
545
31,779
1,275

22,590
14,976
7,044
3,874,296
1,927,160
6,786

BONUS NOW YOU CAN DO IT

Remember that one of the keys to human calculating is *creativity*.

Now, you will also recall that we noticed that numbers can be evenly divided by 30 if they end in 0 and are evenly divisible by 3. The reason for this is that 30 has factors of 3 and 10.

$$3 \times 10 = 30$$

Any number ending in 0 has to have a 10 as one factor. That gives us the "0" requirement for dividing by 30; all that's left is to see whether the number can also be divided by the other factor, 3.

Knowing this, see whether you can formulate your own rule for telling whether a number is evenly divisible by 90.

PART II

PARTS OF NUMBERS

11 Finding Square Roots

LET'S SUPPOSE YOU WANT to cover your old concrete patio with oblong pieces of slate.

The patio, which measures the same on all four sides, covers 144 square feet of space. If each chunk of slate measures one foot long by two feet wide, how many pieces of slate do you need to buy?

There may be easier ways to arrive at the answer than the one we're about to try, but after all, here's an opportunity to use square roots.

Visualize the problem this way:

We know the patio is square, because its sides are all the same length—that's the definition of "square." So picture a square slab of concrete.

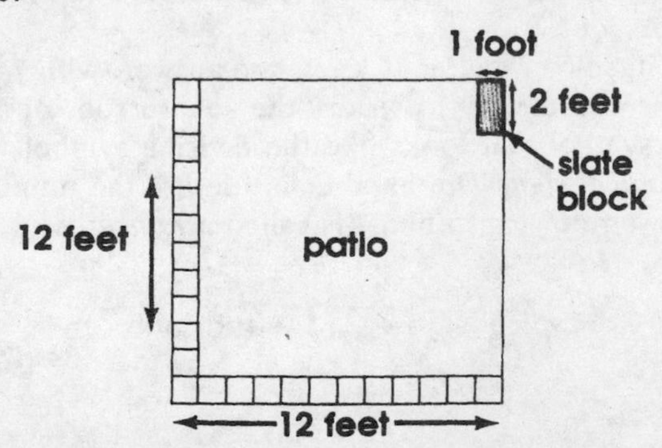

It covers 144 square feet. The first thing we need to know is how many feet it measures on each side. That is, we need to know what number multiplied by itself equals 144. That number will be the same as the number of feet on each side. It's called *the square root of 144*.

When you figure what number multiplied by itself equals another number, you *extract the square root* of the number. Unless you happen to know or happen to have a table of square roots lying around, this procedure is a matter of educated guesswork.

We know that 10 times 10 is 100. Because 144 is larger than 100, we realize that the square root of 144 is larger than 10. It's not 11, because, using our 11s multiplication strategy, we see instantly that 11 times 11 is 121. If we know the 12s table, we recognize that 12 times 12 is 144. That is, *12 is the square root of 144.*

That means the patio is 12 feet long by 12 feet wide. Knowing the square root of 144 makes it easy for us to figure how many one-by-two-foot slate pieces we need. We divide 12 by 1 foot to show how many will fit in one direction—that's 12. Then we divide 12 feet by 2 feet to see how many will fit in the other direction—that's 6. Now multiply 12 × 6 for **the answer: 72.**

Okay, okay! Common sense tells you that if one piece of slate is 1 foot by 2 feet, it contains 2 square feet. So you ought to be able to divide 144 square feet by 2 square feet to get the answer. And yes, that's exactly what you find: 144 square feet ÷ 2 square feet = 72 square feet, the number of slate slabs required to cover the patio. That's the fast way to do it. The slow way shows you *how* it works.

Short of looking up the answer in a table, there's no quick way to extract square roots of larger numbers. But because it's an important procedure, one that you'll need to understand if you go much further in math, we ought to take a look at the strategy for finding square roots.

Let's try it, since we already know the answer, with 144. By the way, a mathematical way to write "the square root of 144" is as $\sqrt{144}$. That symbol that looks like the division symbol with a tail is called a *radical sign.* The number inside it—the number whose square root we're trying to find—is called a *radicand.*

radical sign→ $\sqrt{144}$ ←radicand

First, divide your number (the radicand) into groups of two digits, starting on the right. If the number contains an odd number of digits, then the left-hand group will contain only one digit. If the number contains a decimal point, mark off groups of two figures going in both directions from the point. When counting off from a decimal point, if the last group on the right contains only one figure, add a zero.

1 44 **THINK:** 144 breaks into two groups.

The number of digits in the answer will be the same as the number of groups in your subdivided number.

1 44 **THINK:** 144 has two groups of digits in it. Look at the number formed by the left-hand group and figure, in your head, the largest number whose square is equal to or less than that left-hand number.

1 44 **THINK:** $1 \times 1 = 1$. So, the square root of 1 is 1. This is the first digit in the answer.

I like to write the answer's digits, as I find them, to the right of the number whose square root I'm searching for.

1 44(1

Now, square this first square root and subtract the result from the left-hand number. Bring down the right-hand number next to this sum to form what will be a *partial dividend* for an upcoming step.

```
1  44(1
1
—
0  44
```

THINK: $1^2 = 1$. Subtract that from the left-hand group. Bring down the right-hand group.

Next, double the part of the answer that you've already found. In this case, **1** × 2 = 2. Now, divide this number, known as the *partial divisor*, into the first part of the partial dividend. If there's a remainder, ignore it. The result is the next part of the square root.

```
1 44(12
1
—
2)44
```

THINK: 1 × 2 = 2. Divide 2 into the first part of 44, which is 4: 4 ÷ 2 = **2.** That's the next part of the answer.
Write that digit in the answer.

```
1 44(12
1
—
2)44
```

Now put this same digit (2) next to the first digit of the partial divisor (the other 2).

```
1 44(12
1
—
22)44
```

THINK: Also write the number in the partial divisor. Here, it forms the number 22.

Multiply the partial divisor by the new digit in the answer: 2 × 22 = 44. Subtract that number from the dividend, and, if there are more digits remaining in the original number, draw down the next group. In this case, we've run out of groups, and so we've solved our problem:

1 44(**12** 1 22)44 44 0	**THINK:** $2 \times 22 = 44$. Subtract that from the dividend: $44 - 44 = 0$. There's nothing more to do, and **the answer** is: $\sqrt{144} = $ **12.**

Sometimes this strategy involves some trial and error. Let's try it with a larger number. Find $\sqrt{77,284}$.

7 72 84	**THINK:** Divide the number into groups of two digits. Because we have three groups, the answer will contain three digits.
7 72 84(**2**	**THINK:** What number less than 7 comes closest to being the square root of 7? It must be 2: $2 \times 2 = 4$. $3 \times 3 = 9$, but that's too large.
7 72 84(**2** 4 3	**THINK:** Subtract 2^2 (which is 4) from the left-hand group: $7 - 4 = 3$.
7 72 84(**2** 4 3 72	Now bring down the next group. This will form a *partial dividend:* 372.
7 72 84(**2** 4 4)3 72	Double the part of the square already found: $2 \times 2 = 4$. It forms a partial divisor.
7 72 84(**29** 4 49)3 72	Divide the partial divisor into the first part of the *dividend:* $37 \div 4 = 9$, with a remainder of 1. Disregard the remainder, and write the 9 in the answer: It's the next digit of the square root. Also write the 9 next to the 4 in the partial divisor.

```
   7  72  84(29
   4
49)3  72
   4  41
DOES NOT COMPUTE
```

Now multiply the new divisor by 9: 49 × 9 = 441. But when we try to subtract that from the dividend, we see it is too large. Apparently, the number 9 is too big to fit in this square root. Since 48 × 8 = 384, eight must also be too large. But 7 looks promising. Erase the 9 in our answer and replace it with a 7.

```
   7  72  84(27
   4
47)3  72
   3  29
      43
```

Put the 7 in the divisor and multiply the divisor by 7: 47 × 7 = 329. Subtract this figure from the dividend: 372 − 329 = 43.

```
   7  72  84(27
   4
47)3  72
   3  29
      43  84
```

Bring down the next group of digits, 84, to form a new partial dividend, 4,384.

```
     7  72  84(27
     4
  47)3  72  84(27
     3  29
  54 ) 43   84
```

Once again, double the square root. Use the result as a new partial divisor: 27 × 2 = 54.

```
      7  72  84(278
      4
   47)3  72
      3  29
 548 ) 43   84
```

Now divide 54 into the first part of the dividend: 438 ÷ 54 = 8. Take this as the next digit of the square root. Write it next to the 54 to form the complete divisor.

```
      7  72  84(278
      4
   47)3  72
      3  29
 548 ) 43   84
       43   84
             0
```

Multiply 548 × 8 to get 4,384; subtract that number from the new dividend. The remainder is 0 and **the answer** is **278.**

If you started with a number containing a decimal point, place a decimal point in the root when you reach the two-digit group that contains the decimal point.

This is a cumbersome procedure, no doubt about it. I don't know of any shortcut (other than brute memory) for deriving square roots. But knowing how it works serves three purposes:

1. It helps you understand numbers.
2. It gives you a deep appreciation for the electronic calculator.
3. It keeps you abreast of students in other countries, who do learn how to derive square roots manually—and who are good at it.

The truth is, though, that it's possible to get a quick approximation of an answer. Suppose you wanted to know the approximate square root of 4,923. Look at the first two numbers, remembering their place value: 49(00). Well, you know the square root of 49 is 7. So, the square root of 4,923 is *no less* than 70, because 7(0) × 7(0) = 49(00).

What's the square root of 1,822, approximately? Begin with 18(00). Okay, you know the square root of 16(00) is 4(0), and you know the square of 5(0) is 25(00). Now 18(00) falls between 16(00) and 25(00), and so its square root must fall somewhere between 4(0) and 5(0). That is, the square root of 1,822 is somewhere between 40 and 50. Because 18(00) is closer to 16(00) than to 25(00), the answer must be closer to 40 than to 50. It's probably less than 45.

The exact square root of 4,923 is 70.16, and that of 1,822 is 42.68. Our estimates are not far off the mark!

In olden days before the invention of the calculator, people compiled tables of square roots. I have included such a table at the end of this book. So, you have three ways to find a square root: figure it out with your own mind, look it up, or punch it into a machine.

12 Finding Cube Roots

DERIVING CUBE ROOTS, an even more complicated process than finding square roots, lends itself to human calculating. Anyone who knows how involved it is to find the square root of a number—much less a cube root—will be startled when they see you perform one of these tricks.

First, let me explain that a cube is the product of a number multiplied by itself twice. The cube root is that number multiplied to form the product or cube. The cube root of 27, for example, is 3:

$$3 \times 3 \times 3 = 27 \quad (3 \times 3 = 9;\ 9 \times 3 = 27)$$

You write the expression "cube root of 27" with a radical, the same kind of symbol used with the square root, except that you put a little *3* inside the crook of the symbol:

$$\text{root index}\rightarrow\sqrt[3]{27} = 3\leftarrow\text{cube root}$$

radical symbol↗ ↑ radicand

That little 3 is called the *root index*.

When you cube a number, as we just suggested, you multiply it by itself twice. Indicate this process with a raised *3* after the number to be cubed: $3^3 = 27$ ("three cubed equals twenty-seven)." You may find it easier to remember this relationship by recalling that you write the number down *three* times when you multiply to find a cube.

You can visualize cubes and cube roots by thinking of a stack of blocks. Each block is the same size on all sides, and the stack is itself the same size on all sides—in other words.

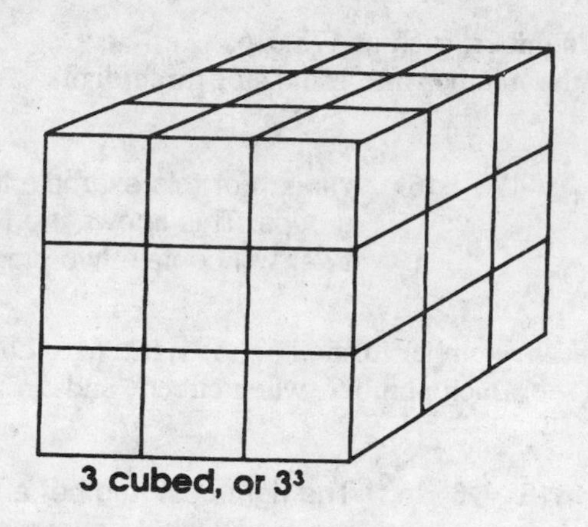

3 cubed, or 3^3

CUBE ROOTS THE EASY WAY

To pull off this strategy, you need to know the following chart:

$1^3 = 1$
$2^3 = 8$
$3^3 = 27$
$4^3 = 64$
$5^3 = 125$
$6^3 = 216$
$7^3 = 343$
$8^3 = 512$
$9^3 = 729$

Let's take a look at those answers again:

1 8 27 64 125 216 343 512 729

Notice that each answer ends in a different digit. This means that, when whole numbers from 1 to 9 are cubed, *no answer will end in the same digit!*

Here's how this little gem of information works for human calculators:

Take a number, such as 175,616.

Divide the number into groups of three digits:

175 616 — THINK: For this example there are two groups. This shows me that the answer will contain two digits.

Now look at the number formed on the right: 616. Look at the chart above and see which number, when cubed, ends in a 6.

$\sqrt[3]{175\ 616} = \mathbf{6}$ — The rightmost number ends in 6. Of the cubed whole numbers 1–9, only 6^3 has an answer that ends in 6: 216. That 6^3 means the second half of the answer is **6.**

Next, look at the number formed by the left-hand set of three digits: 175. Ask yourself, between which two numbers does the number 175 fit? You'll find those two numbers on the cube chart above. The answer is, of course, between 125 and 216. Now, you *always* use the *smaller* of the two numbers that the left-hand number falls between—in this case, 125. The cube root of 125 is 5, and so the first digit in our answer is **5.**

$\sqrt[3]{175\ 616} = \mathbf{56}$ **THINK:** Where does 175 fall among the answers in the cube-root chart? It falls between 125 and 216. Because 125 is the smaller of those two, the first part of the answer is the cube root of 125: **5.**

The answer, then, is **56.** To test it, multiply 56 by itself three times:

$$\begin{array}{r} 56 \\ \times\ 56 \\ \hline 3{,}136 \end{array} \qquad \begin{array}{r} 3{,}136 \\ \times\ 56 \\ \hline 175{,}616 \end{array}$$

Yes, 56^3 is 175,616, and so the opposite, $\sqrt[3]{175{,}616}$, is 56.
This strategy works on numbers to one million.

Now You Can Do It

Remember, divide these numbers into groups of three, starting from the right—just as you divided numbers for square roots into groups of two. Practice your prestidigitation well and you will stun your teachers!

$\sqrt[3]{704{,}969}$ $\sqrt[3]{778{,}688}$ $\sqrt[3]{85{,}184}$ $\sqrt[3]{5{,}832}$

$\sqrt[3]{287{,}496}$ $\sqrt[3]{753{,}571}$ $\sqrt[3]{912{,}673}$ $\sqrt[3]{4{,}913}$

Do you want to know *why* your teachers are stunned when they see you derive a cube root in your head? Because they learned to find cube roots in this way:

THE LONG WAY TO DERIVE A CUBE ROOT

Let's try this on one of the numbers whose cube root you've found the easy way: 704,969.

First, divide the number into three-digit groups, exactly as above:

$\sqrt[3]{704\ 969} =$	**THINK:** Two groups means the answer contains two digits.
$\sqrt[3]{704\ 969} = \mathbf{8}$	Find the greatest cube that fits into the left-hand group. From the chart, you know the number 704 falls between 512 and 729; 729 is too big, so the closest perfect cube to 704 must be 512. Write its cube root, which is **8,** as the first digit of the answer.
$\sqrt[3]{704\ 969} = \mathbf{8}$ $\underline{512}$ 192	Next, write that figure's cube below the left-hand group and subtract.
$\sqrt[3]{704\ 969} = \mathbf{8}$ $\underline{512}$ 192 969	Bring down the next group to form a partial remainder.
$64 \times 300 = 19{,}200$	Now multiply the square of the first figure in the root ($8 \times 8 = 64$) times 300.
$\sqrt[3]{704\ 969} = \mathbf{8}10$ $\underline{512}$ 19,200)192 969	Divide the partial dividend by this answer. Write the answer as a trial number for the next digit in the cube root.
$\sqrt[3]{704\ 969} = \mathbf{8}10$ $\underline{512}$ 19,200)192 969 540 $\underline{100}$ 19,840	Next, add the first digit and the trial answer: $8 + 10 = 18$. Multiply this product by 30: $18 \times 30 = 540$. Then, square the trial answer: $10^2 = 100$. Add these two numbers to the trial divisor: $19{,}200 + 540 + 100 = 19{,}840$. This is supposed to be the complete divisor.

19,840 × 10 = 198,400	Now multiply the new complete divisor by the trial figure in your answer: 19,840 × 10 = 198,400. If this number is smaller than the dividend, subtract it from dividend. But if it is larger than the dividend, you have to start over with a new trial divisor!

$$\begin{array}{r} \sqrt[3]{704\ 969} = 810 \\ \underline{512} \\ 19{,}200)192\ 969 \\ 540 \\ \underline{100} \\ 19{,}840)\underline{198{,}400} \end{array}$$

DOES NOT COMPUTE

$\sqrt[3]{704\ 969} = 89$ $\underline{512}$ 19,200)192 969	Sigh! Is it any wonder arithmetic gave your mother conniption fits when she was in grade school? Try the next smaller number, 9.
8 + 9 = 17	Once again, add the two digits.
17 × 30 = 510	Multiply the result by 30.
$9^2 = 81$	Square the trial answer: 9 × 9 = 81.
19,200 510 $\underline{81}$ 19,791	Add the results of the last two steps to the partial divisor: 19,200 + 510 + 81 = 19,791. This is the proposed new complete divisor.
19,791 × 9 = 178,119	Multiply this complete divisor by the trial answer to get 178,119. This number is smaller than the dividend. At last, you can subtract.
$\sqrt[3]{704\ 969} = 89$ $\underline{512}$ 19,200)192 969 $\underline{-\ 178{,}119}$ 14,850	**The answer** is **89.** To test it, multiply 89 × 89 × 89: 89^3 = 704,969.

You persist with this exercise until you run out of groups of figures—or, more likely, until you run out of steam. If you need to derive the cube root of a number with a decimal point in it, you mark off your groups of three in each direction, starting from the left and from the right of the decimal point; if the right-hand group has fewer than three digits, add zeros to fill. The answer has as many numbers to the left of the decimal point as there are groups to the left of the point in the original number.

Our parents and grandparents were subjected to this awful procedure in grade school. Today's teachers are either more enlightened or less persistent than teachers of yesteryear—it's hard to find a modern arithmetic textbook that explains how to derive cube roots.

If you have to find the cube root of a number larger than six figures, look it up in a table. I'm including a table of cube roots at the end of this book.

USING THE ELECTRONIC CALCULATOR TO DERIVE HIGHER ROOTS

A square is the product of a number multiplied by itself.

$$3 \times 3 = 9; \sqrt{9} = 3$$

Or, to put it another way, the *second power* of 3 is 9. Remember, you write a number down two times when you multiply to find the second power.

A cube is the product of a number multiplied by itself twice. A cube root is the number that is multiplied to form the product or cube.

$$3 \times 3 \times 3 = 27; 3^3 = 27$$

$$\sqrt[3]{27} = 3$$

That is, three cubed is 27, or, the *third power* of 3 is 27. The cube root of 27 is 3.

Now, a *fourth root* is a number multiplied by itself four times to form a new number:

$$3 \times 3 \times 3 \times 3 = 81;\ 3^4 = 81$$
$$\sqrt[4]{81} = 3$$

Or, the *fourth power* of 3 is 81. The fourth root of 81 is 3. You write a number down four times when multiplying to the fourth power.

Similarly, the *fifth power* of 3 (3^5) is the same as $3 \times 3 \times 3 \times 3 \times 3 = 243$. The *fifth root* of 243 ($\sqrt[5]{243}$) is 3.

So it goes. You could proceed with these *higher powers* indefinitely. For example, 4^9 is the same as $4 \times 4 \times 4 \times 4 \times 4 \times 4 \times 4 \times 4 \times 4$. And of course, as we all recognize, $4^9 = 262{,}144$. Which is to say, $\sqrt[9]{262{,}144} = 4$.

Many higher roots can be found by combining square and cube root extractions. (Too bad the word for this sounds like dental surgery—although, given what our parents had to go through in school, it's not surprising.) Consider: The *fourth root* is the square root of the square root. Take the square root of 81, which is 9; now get the square root of that, which is 3, and you have the fourth root of 81.

It's easy to perform this on a calculator. Most calculators have a square root button—it's usually marked with a radical sign ($\sqrt{\ }$). To use it, you simply enter the number whose square root you want to derive and then press the square root button. Every time you enter "square root," the calculator derives the square root of the number shown in the display. This allows you to derive fourth roots and eighth roots very simply.

To find the fourth root of 531,441, for example, you would first find its square root:

$$\sqrt{531{,}441} = 729$$

Then you would find the square root of 729, which is 27. And if you worked it out the long way, you would find that

$$\sqrt[4]{531{,}441} = 27$$

Other rules of thumb for extracting higher roots:

The *sixth* root is the cube root of the square root. (Find the square root on your calculator and then do the cube root the Human Calculator's way.)

The *eighth* root is the square root of the fourth root, which is the same as the square root of the square root of the square root. (Push the square root button on your calculator three times.)

The *ninth* root is the cube root of the cube root.

Now You Can Do It

Some of us may regard this as the Ultimate Challenge. However, knowing what you do now, you should find it relatively easy (say, compared with the way your grandmother found it) to derive these roots:

$\sqrt[3]{21{,}952}$ $\sqrt[3]{185{,}193}$ $\sqrt[3]{941{,}192}$

$\sqrt[4]{256}$ $\sqrt[4]{194{,}481}$ $\sqrt[4]{57{,}289{,}761}$

$\sqrt[6]{46{,}656}$ $\sqrt[6]{1{,}771{,}561}$ $\sqrt[6]{1{,}073{,}741{,}824}$

$\sqrt[9]{512}$

13 Fractions and Decimals

A FRACTION IS SIMPLY a part of a whole number. Look at a ruler, for example, and examine the way an inch is broken into smaller parts: it shows the inch has two halves, four quarters, eight eighths, and sixteen sixteenths. These fractions are expressed like this: $\frac{1}{2}$; $\frac{1}{4}$; $\frac{1}{8}$; $\frac{1}{16}$. Sometimes the horizontal line is replaced with a slash separating the two figures: 1/2; 1/4; 1/8; 1/16.

These numbers are called *common fractions*. When you write them with a whole number above a line and a whole number below the line, the top number is called the *numerator* and the bottom number is the *denominator*. Derived from Latin, the word "denominator" means "this is the number we use to name the fraction." In $\frac{3}{4}$, for example, we know we're talking about quarters of some object or quantity. "Numerator" is another Latinate term meaning "this is the number we use to count the fractions." In $\frac{3}{4}$, we're saying that we have three of our quarter things. The numerator and the denominator are sometimes called the *terms* of a fraction.

$$\frac{3}{4} \quad \begin{matrix}\underline{\textbf{numerator}} \\ \textbf{denominator}\end{matrix}$$

A fraction represents a kind of division. Until we get further into our discussion, keep this in mind: The line in a fraction means that you divide the upper number (numerator) by the lower number (denominator). It has the same meaning as a division sign. That is:

$$\frac{3}{4} = 3 \div 4 = 4\overline{)3}$$

When a whole number and a fraction occur together, you have a *mixed number*. Measure off one and one-half inches on your ruler, for example, and you'll have a mixed number: $1\frac{1}{2}$ inches.

In order to work with mixed numbers, you have to convert them into common fractions. There are two ways to do this. One is by converting the whole number into a fraction with the same denominator as the fraction that accompanies it. Then add the new numerator to the numerator of the accompanying fraction to change the mixed number into a fraction. You can do this by remembering that any fraction with the same numerator and denominator equals 1 provided that the numerator and denominator are not 0s.

$$\frac{1}{1} = 1 \qquad \frac{2}{2} = 1 \qquad \frac{5}{5} = 1 \qquad \frac{7{,}289}{7{,}289} = 1$$

If we try this on $1\frac{1}{2}$, we see that:

$$1 = \frac{2}{2}$$

$$1\frac{1}{2} = \frac{2}{2} + \frac{1}{2} = \frac{3}{2}$$

You also can convert a mixed number into a common fraction by multiplying the whole number by the fraction's denominator and then adding the result to the numerator. Like this:

$1\frac{1}{2}$	**THINK:** $2 \times 1 = \mathbf{2}$
$\frac{2 + 1}{2}$	Add 2 to the numerator, which is 1, to get 3.
$\frac{3}{2}$	$1\frac{1}{2} = \frac{3}{2}$

Now You Can Do It

Change these mixed numbers into fractions:

$2\frac{3}{4}$ $3\frac{4}{5}$ $10\frac{3}{12}$ $9\frac{6}{8}$ $16\frac{13}{32}$

You can add, subtract, multiply, and divide fractions and whole numbers, but there are a few tricks you need to know first.

HOW TO ADD FRACTIONS

Before you can add or subtract fractions, you have to change them so they have the same denominator. This is called *finding a common denominator.* You can find the common denominator of two or more fractions several ways.

Suppose, for example, that you wanted to add $\frac{1}{4}$ and $\frac{1}{2}$.

If one fraction's denominator divides evenly into the second fraction's denominator, multiply *both* terms of the first fraction by the number required to make the first fraction's denominator the same as the second fraction's denominator.

Suppose we want to add $\frac{1}{2} + \frac{1}{4}$.

$$\frac{1}{2} + \frac{1}{4}$$

THINK: The denominator *2* goes 2 times into the denominator *4.* So if we multiply both terms of $\frac{1}{2}$ by 2, we can convert the half into fourths.

$$\frac{1 \times 2}{2 \times 2} = \frac{2}{4}$$

$\frac{1}{2} = \frac{2}{4}$. Now we can add fourths.

$$\frac{2}{4} + \frac{1}{4}$$

The converted problem is $\frac{2}{4} + \frac{1}{4}$. Just add the numerators: $1 + 2 = 3$. Place that number over the denominator of 4.

$$\frac{3}{4}$$

The answer is $\frac{3}{4}$.

Another way to accomplish the same goal—more complicated, but useful with larger numbers—is to find a common denominator by multiplying both denominators together. Using the same example, multiply the denominator 2 by the denominator 4 to find the common denominator, 8:

$$2 \times 4 = 8$$

Now change each fraction to eighths:

$$\frac{1}{2} = \frac{4}{8}$$

How? Multiply both terms by the number required to change the denominator to an 8. In this case, the number is 4:

$$\frac{1 \times 4 = 4}{2 \times 4 = 8}$$

$$\frac{1}{4} = \frac{2}{8}$$

The number required to change this denominator to an 8 is 2. Multiply both terms by 2:

$$\frac{1 \times 2 = 2}{4 \times 2 = 8}$$

$$\frac{4}{8} + \frac{2}{8} = \frac{6}{8}$$

Add the eighths.

Notice that 6 and 8 are both divisible by 2. Divide them to reduce the fraction.

$$\frac{6 \div 2 = 3}{8 \div 2 = 4}$$

The answer is $\frac{3}{4}$.

That maneuver—dividing both parts of a fraction by the same number—is called *reducing the fraction.* It allows you to turn a ridiculous-looking number, such as $\frac{45}{120}$, into a more understandable number:

$$\frac{45 \div 15 = 3}{120 \div 15 = 8}$$

The fraction $\frac{45}{120}$ is said to have the same *value* as $\frac{3}{8}$. They're really just two ways of saying the same thing.

The secret behind this is the same that works in some of our division strategies: You don't change the answer (or the "value" of a fraction) when you multiply or divide *both parts* of the division problem—or the fraction—by *the same number.*

Now You Can Do It

Make these fractions have common denominators:

$$\frac{2}{5} \text{ and } \frac{4}{8}$$

$$\frac{3}{4} \text{ and } \frac{1}{2}$$

$$\frac{7}{10} \text{ and } \frac{2}{15}$$

$$\frac{12}{18} \text{ and } \frac{4}{5}$$

$$\frac{2}{3} \text{ and } \frac{4}{5} \text{ and } \frac{3}{4}$$

$$\frac{6}{10} \text{ and } \frac{9}{11} \text{ and } \frac{5}{6}$$

WEIRD-LOOKING FRACTIONS

When a fraction's numerator is the same as or larger than its denominator, it's called an *improper* fraction. Any fraction that is not reduced to its lowest number or is not written in its simplest form is an improper fraction. $\frac{173}{120}$ is an improper fraction. So are $\frac{7}{7}$, $\frac{81}{72}$, and $\frac{4}{3}$. Ordinary, normal-looking fractions whose numerators are smaller than their denominators are called *proper* fractions. How do you bring an improper fraction back into the fold of propriety? Always *divide* and conquer!

$$\frac{7}{7} = 7 \div 7 = 1$$

Remember, any time a fraction's numerator and denominator are the same, the fraction equals 1.

$$\frac{81}{72} = 72 \div 81 = 1\,\frac{9}{72}$$

Nine seventy-seconds? What does that mean? Notice that nine goes evenly into 72. Divide both parts by 9 to reduce the fraction:

$$\frac{9 \div 9}{72 \div 9} = \frac{1}{8}$$

So, $1\frac{9}{72} = 1\frac{1}{8}$.

To reduce a fraction like $^9/_{15}$, divide both terms by a number that will go into both. We see that 9 is divisible by 3, and so is 15. Fifteen is also evenly divisible by 5, but 5 doesn't go into 9. So let's divide both parts by 3:

$$\frac{9 \div 3}{15 \div 3} = \frac{3}{5}$$

Some fractions, such as $\frac{9}{10}$ or $\frac{53}{120}$, cannot be reduced. Use the strategies for figuring divisibility, learned in Chapter 10, to help you see whether the terms of a fraction can be evenly divided by the same number. We know that 53 divides by 1 and by 53, which doesn't do us much good, since we also can see that 120 does not divide evenly by 53. Dividing both parts by 1, of course, changes nothing. Although 120 can divide by lots of factors, none of them go evenly into 53. So, we're stuck with trying to imagine $\frac{53}{120}$ as a real quantity.

I wonder how much $\frac{53}{120}$ is. Does this crazy-looking figure have any meaning?

If I divided a huge pizza into 120 pieces, half of it would be 60 pieces. Fifty-three pieces would be seven pieces less than half the pizza, so $\frac{53}{120}$ must be a little less than $\frac{1}{2}$. A third of the pizza would contain 40 pieces ($120 \div 3 = 40$). Fifty-three is 13 more than 40, so $\frac{53}{120}$ must be a fair amount more than $\frac{1}{3}$. By visualizing this number as part of a real object, I can estimate quickly that $\frac{53}{120}$ is closer to $\frac{1}{2}$ than $\frac{1}{3}$.

What are we doing here? We're using the fifth key to human calculating—creativity—to help us understand what numbers mean. Never be shy about using something different to help you follow math. If it works, use it!

Now You Can Do It

Add these fractions. If some of them result in improper fractions, change the fractions to mixed numbers.

$\frac{1}{4} + \frac{1}{8}$ $\quad$ $\frac{1}{8} + \frac{2}{3}$ $\quad$ $\frac{5}{6} + \frac{8}{9}$

$\frac{4}{5} + \frac{7}{12}$ $\quad$ $\frac{9}{10} + \frac{7}{8}$ $\quad$ $\frac{10}{45} + \frac{5}{8}$

$\frac{15}{16} + \frac{3}{20}$

ADDING FRACTIONS THE HUMAN CALCULATOR'S WAY

Here's a fast strategy to add any two fractions:

First, multiply the denominators. Then, cross-multiply the numerators and denominators and add. Let's try it:

What is $\frac{1}{3} + \frac{1}{2}$?

$\frac{1}{\mathbf{3}} \times \frac{1}{\mathbf{2}} = 6$ — **THINK:** Multiply the denominators: 3 × 2 = 6. This is a new denominator.

$\frac{1}{} \times \frac{}{2} = 2$ — Cross-multiply the first numerator times the second denominator: 1 × 2 = 2.

$\frac{}{3} \times \frac{1}{} = 3$ — Now cross-multiply the other numerator and denominator: 3 × 1 = 3. Add these two products: 2 + 3 = 5. This is the new numerator.

$\frac{2 + 3}{6} = \frac{\mathbf{5}}{\mathbf{6}}$ — **The answer is $\frac{5}{6}$.**

Astonishingly simple, isn't it? Maybe that example was too easy. Let's try $\frac{4}{5} + \frac{2}{3}$.

$\frac{4}{5} \times \frac{2}{3} = 15$	THINK: Multiply the denominators to get the new denominator: 15.
$\frac{4}{\ } \times \frac{\ }{3} = 12$	Multiply the first numerator times the second denominator: 4 × 3 = 12.
$\frac{\ }{5} \times \frac{2}{\ } = 10$	Multiply the second numerator times the first denominator: 2 × 5 = 10.
$\frac{12 + 10}{15} = \frac{22}{15}$	Add the two figures for the new numerator and put it over the new denominator.
$22 \div 15 = 1$, r. 7	To reduce this improper fraction to a mixed fraction, divide 22 by 15. The result is 1 with a remainder of 7. Place the 7 over the divisor to
1, r. 7 = $1\frac{7}{15}$	get the fraction in **the answer: $1\frac{7}{15}$.**

One other magic trick in this department: If you get to add two fractions whose numerators are 1, just put the sum of the denominators over the product of the denominators. Like this:

$\frac{1}{3} + \frac{1}{8}$	THINK: Add the denominators 3 + 8 = 11.
$3 + 8 = 11$	This is the new numerator.
$3 \times 8 = 24$	Next, multiply the denominators: 3 × 8 = 24. This is the new denominator.
$\frac{1}{3} + \frac{1}{8} = \frac{11}{24}$	**The answer** is $\frac{11}{24}$.

Now You Can Do It

Try adding these fractions the easy way:

$$\frac{3}{4} + \frac{5}{9} \qquad \frac{1}{6} + \frac{4}{5} \qquad \frac{14}{15} + \frac{5}{8}$$

$$6\frac{3}{5} + \frac{7}{8} \qquad 10\frac{5}{6} + \frac{7}{9} \qquad \frac{51}{65} + \frac{2}{3}$$

$$\frac{1}{3} + \frac{1}{2} \qquad \frac{1}{8} + \frac{1}{9} \qquad \frac{1}{90} + \frac{1}{360}$$

HOW TO SUBTRACT FRACTIONS

Subtracting with fractions works the same way as addition. Find a common denominator. Change all the fractions to new fractions that have that common denominator. Subtract the smaller numerator from the larger. And, if necessary, reduce the resulting fraction to understandable terms.

Let's start with a couple of fractions that have the same denominator:

$$\frac{112}{143} - \frac{96}{143} = \frac{16}{143}$$

Pretty easy: just subtract 96 from 112. Put the result, 16, over the denominator (143) for the answer: $\frac{16}{143}$.

And if we wanted to subtract $\frac{1}{3}$ from $\frac{2}{5}$?

$$\frac{2}{5}$$

$$-\frac{1}{3}$$

THINK: What's a common denominator for $\frac{2}{5}$ and $\frac{1}{3}$? Multiply the denominators: $3 \times 5 = 15$.

$$\frac{2 \times 3}{5 \times 3} = \frac{6}{15}$$

How to change $\frac{2}{5}$ into fifteenths? Multiply both parts by the number required to change 5 to 15. That number is 3. So, multiply 2×3 and 5×3.

$$\frac{1}{3} \times \frac{5}{5} = \frac{5}{15}$$

Okay. Now change $\frac{1}{3}$ into fifteenths. Do the same thing—multiply by the number required to change the denominator (3) into 15. That number is 5. Multiply 1×5 and 3×5.

$$\begin{array}{r} \frac{6}{15} \\ -\ \frac{5}{15} \\ \hline \frac{1}{15} \end{array}$$

Now you can subtract apples from apples. Subtract 5 from 6 to get 1, the answer's denominator.

The answer is $\frac{1}{15}$.

Now You Can Do It

Don't forget to reduce the answer to its lowest terms whenever possible.

$$\frac{3}{5} - \frac{1}{5} \qquad \frac{4}{5} - \frac{3}{4} \qquad \frac{3}{4} - \frac{5}{8}$$

$$\frac{2}{5} - \frac{1}{8} \qquad \frac{9}{11} - \frac{7}{22}$$

ADDING AND SUBTRACTING MIXED NUMBERS

When you have to work with mixed numbers, convert the mixed numbers into fractions and proceed with the same steps we just went through.

For example: I have a bowl that holds $3\frac{3}{4}$ cups. My cake recipe calls for a total of $2\frac{1}{2}$ cups of flour, sugar, and baking powder, plus a total of $1\frac{5}{8}$ cups of milk, shortening, and vanilla. Can I use the bowl to mix up this cake?

To know that, I have to know how much $2\frac{1}{2}$ cups and $1\frac{5}{8}$ cups of ingredients adds up to.

Remember: To change mixed numbers into fractions, multiply

the whole number by the denominator and add the product to the numerator.

$$\begin{array}{r} 2\frac{1}{2} \\ +\ 1\frac{5}{8} \\ \hline \end{array}$$

THINK: $2\frac{1}{2}$ expressed as a fraction equals $\frac{5}{2}$:

$$2 = \frac{4}{2}$$

$$\frac{4}{2} + \frac{1}{2} = \frac{5}{2}$$

$1\frac{5}{8}$ equals $\frac{13}{8}$:

$$1 = \frac{8}{8}$$

$$\frac{8}{8} + \frac{5}{8} = \frac{13}{8}$$

So far, so good. Now, to get the new numerator, we will multiply the *old* numerators together, then multiply the *old* denominators together, and then add the products. The result will be the *new numerator*.

$$\frac{5}{2} \times \frac{13}{8} = 16$$

Warp 9, Mr. Sulu! Multiply 2 × 8 to get a new denominator: 16.

$$\frac{5}{\ } \times \frac{\ }{8} = 40$$

Cross-multiply the old numerators and the old denominators. 5 × 8 = 40; and

$$\frac{\ }{2} \times \frac{13}{\ } = 26$$

2 × 13 = 26.

$$40 + 26 = 66$$

$$\frac{40 + 26}{16} \quad \frac{66}{16}$$

$$16\overline{)66} \quad \underline{4,\ r.\ 2}$$

$$4,\ r.\ 2 = 4\tfrac{2}{16}$$

$$\frac{2 \div 2 = 1}{16 \div 2 = 8}$$

$$4\frac{2}{16} = 4\frac{1}{8}$$

Add the results to get 66. Put it over the new denominator, which was 16. Reduce this improper fraction to a mixed number by dividing 16 into 66. The result is 4 with a remainder of 2, which is the same as $4\frac{2}{16}$. Now reduce $\frac{2}{16}$ to a more manageable fraction by dividing both parts by 2, and you get **the answer:** $4\frac{1}{8}$.

Now You Can Do It

Find the answers to these puzzles:

$$9\frac{2}{7} - 5\frac{3}{4} \qquad 7\frac{3}{5} + 3\frac{1}{10}$$

$$8\frac{5}{6} - 5\frac{1}{12} \qquad \frac{11}{3} + 3\frac{9}{55}$$

$$2 + 3\frac{7}{8} + 5\frac{1}{5} - 9\frac{5}{16}$$

Well, $4\frac{1}{8}$ cups is pretty clearly more than my $3\frac{3}{4}$-cup bowl will hold. Unless I have a larger bowl, I'll have to reduce my recipe to about $\frac{2}{3}$ of the original.

MULTIPLYING BY FRACTIONS

Wouldn't you know it? I don't have a larger bowl. Guess I'll have to cut my recipe down. So what's $\frac{2}{3}$ of $4\frac{1}{8}$ cups?

To find a fraction of a larger number, you *multiply* the number by the fraction, instead of dividing. That means I need to multiply $4\frac{1}{8} \times \frac{2}{3}$.

Multiplying fractions is easy. You just multiply the numerators together and then multiply the denominators together. You can sometimes simplify things even further by reducing one or more of the fractions to lower terms, or by performing a strategy called *canceling,* which we'll describe below.

Now, $4\frac{1}{8}$ is a mixed number. To multiply it by $\frac{2}{3}$, first convert it to a fraction.

$$4\frac{1}{8} = \frac{33}{8}$$

Now, multiply numerators and denominators:

$$\frac{33 \times 2}{8 \times 3} = \frac{66}{24}$$

$$\frac{66 \div 6}{24 \div 6} = \frac{11}{4}$$

$$11 \div 4 = 2, \text{r. } 3$$

$$\frac{11}{4} = \mathbf{2\frac{3}{4}}$$

THINK: multiply the top and bottom terms. The result, $\frac{66}{24}$, can easily be reduced by dividing both terms by 6: Now convert this simplified answer, $\frac{11}{4}$, to a mixed number by dividing 11 by 4. The result is 2 with a remainder of 3.
Place the 3 over the divisor of 4 to find **the answer, $2\frac{3}{4}$ cups.**

Two and three-quarter cups will fit handily inside my bowl, which holds $3\frac{3}{4}$ cups. I'll have to multiply the amount of each ingredient by $\frac{2}{3}$, and the cake will only be $\frac{2}{3}$ as large as it might have been.

What if you want to multiply a whole number by a fraction? You have two choices, and you can use whichever is easier. You could (1) multiply the whole number by the numerator and then divide the product by the denominator, or (2) divide the number first and then multiply that result by the numerator. Either strategy works; choose the simpler path.

Strategy 1

$$15 \times \frac{3}{5}$$

$$15 \times 3 = 45$$

$$5\overline{)45} = 9$$

$$15 \times \frac{3}{5} = \mathbf{9}$$

THINK: Multiply the whole number by the fraction's numerator. $15 \times 3 = 45$. Now divide that by the denominator: $45 \div 5 =$ **9**. And that's **the answer.**

Strategy 2

$$15 \times \frac{3}{5}$$

$$5\overline{)15} = 3$$

THINK: Divide the whole number by the denominator: $15 \div 5 = 3$.

$$3 \times 3 = 9$$

$$15 \times \frac{3}{5} = \mathbf{9}$$

Now multiply by the numerator: $3 \times 3 =$ **9,** which is the same **answer** we found above.

Now here's a neat little activity known as *canceling* or *cancellation.* It lets you reduce the terms in a bunch of fractions to smaller, more manageable figures. Sometimes you can cancel numbers out altogether! Then you don't have to deal with them at all.

Suppose you want to multiply

$$\frac{2}{3} \times \frac{3}{5}$$

See those two 3s? They cancel each other out. Three divided into 3 equals 1; $1 \times 5 = 5$ and $1 \times 2 = 2$. Those 3s might as well not be there. Just cross them out. All you have left now is a 2 in the numerator and a 5 in the denominator. And yes, the answer to $\frac{2}{3} \times \frac{3}{5}$ is $\frac{2}{5}$.

$$\frac{2}{{}_1\cancel{3}} \times \frac{\cancel{3}^1}{5} = \frac{2}{5}$$

Cross-divide the numbers in the numerators and denominators, and then replace the originals with the new numbers. Like this:

$$\frac{{}^1\cancel{2}}{4} \times \frac{8}{\cancel{12}_6}$$

THINK: Two goes into 12 six times; cancel the 2 and turn the 12 into a 6.

$$\frac{{}^1\cancel{2}}{{}_1\cancel{4}} \times \frac{\cancel{8}^2}{\cancel{12}_6}$$

Four goes into 8 two times; cancel the 4 and turn the 8 into a 2. Now multiply by the reduced terms:

$$\frac{1}{1} \times \frac{2}{6} = \frac{2}{6} = \frac{1}{3}$$

The result, $\frac{2}{6}$, reduces to **the answer:** $\frac{1}{3}$.

If you have fractions whose terms end in 0s, you can cancel out the same number of 0s from numerators and denominators.

$$\frac{4}{10\cancel{0}} \times \frac{2\cancel{0}}{45}$$

$$\frac{4}{10} \times \frac{2}{45}$$

THINK: Cancel a 0 from the 20 and also from the 100—that is, from a numerator and from a denominator.

$$4 \times 2 = 8$$

$$10 \times 45 = 450$$

Multiply the new terms, top by top and bottom by bottom.

$$\frac{8}{450} = \frac{4}{225}$$

Reduce the fraction by dividing both terms by 2: $\frac{8}{450} = \frac{4}{225}$, **the answer.**

$$\frac{4}{100} \times \frac{20}{45} = \frac{\mathbf{4}}{\mathbf{225}}$$

You can get completely carried away with this procedure and end up simplifying what looks like a very complicated problem. For example:

$$\frac{{}^{1}\cancel{2}}{4} \times \frac{{}^{1}\cancel{5}}{\cancel{8}_{4}} \times \frac{{}^{1}\cancel{16}}{4\cancel{0}_{8}} \times \frac{3\cancel{0}}{3\cancel{2}_{2}} \times \frac{15}{20} = \frac{1}{4} \times \frac{1}{4} \times \frac{1}{8} \times \frac{3}{2} \times \frac{15}{2} = \frac{45}{512}$$

$$72000/819200 = 720/8192 = 45/512$$

When a numerator and a denominator each have numbers that end in 0, you can cancel as many 0s as you find in the number that has the fewest 0s. For example:

$$\frac{1\cancel{00}}{250\cancel{00}} = \frac{1}{250} \qquad \frac{460\cancel{00}}{85\cancel{00}} = \frac{460}{85} = 5\frac{{}^{7}\cancel{5}}{17}$$

Of course, this can come in extremely handy when you're multiplying:

$$\frac{1}{10} \times \frac{250}{365} = \frac{25}{365} \qquad \frac{14}{100} \times \frac{180}{13} = \frac{252}{130}$$

Put it all together, and you arrive at a shortcut like this:

$$\frac{{}^{2}\not{14}}{100} \times \frac{180}{7} = \frac{2 \times 18}{70} = \frac{36}{70} = \frac{18}{35}$$

Now You Can Do It

$$\frac{1}{2} \times \frac{2}{3} \qquad \frac{3}{7} \times \frac{5}{6} \qquad \frac{16}{21} \times \frac{5}{8}$$

$$\frac{10}{27} \times \frac{97}{100} \qquad \frac{1}{4} \times \frac{2}{3} \times \frac{3}{5}$$

$$\frac{7}{10} \times \frac{5}{21} \times \frac{12}{13}$$

DIVIDING BY FRACTIONS

To divide a number (any number, including another fraction), invert the divisor and follow the same steps you use in multiplication.

Let's try dividing $\frac{6}{7}$ by $\frac{1}{2}$.

$\frac{6}{7} \div \frac{1}{2}$ — **THINK:** Invert the divisor, which is $\frac{1}{2}$. It becomes $\frac{2}{1}$.

$\frac{6}{7} \times \frac{2}{1}$ — Now multiply the new numerators and the new denominators: $6 \times 2 = 12$; $7 \times 1 = 7$.

$7 \times 1 = 7$ — The new fraction is $\frac{12}{7}$.

$\frac{12}{7} = 1 \text{ r. } 5$ — Reduce this improper fraction to **$1\frac{5}{7}$, the answer.**

Very simple. Now try something a little more complicated:

$12 \div \frac{3}{4}$ — Invert the divisor: $\frac{3}{4}$ becomes $\frac{4}{3}$.

$12 \times \frac{4}{3} = \frac{48}{3}$ — Proceed as for multiplication: $12 \times 4 = 48$. Reduce the improper fraction: $\frac{48}{3}$ = **16, the answer.**

If the divisor is a whole number and the dividend is a fraction, invent a fraction for the divisor by placing the whole number over 1 (any number divided by 1 is itself; since a fraction is an expression of division, dividing by 1 makes a fraction with a denominator of 1: $4 = 4 \div 1 = \frac{4}{1}$).

$\frac{3}{4} \div 4$

$\frac{3}{4} \times \frac{1}{4}$

THINK: 4 is the same as $\frac{4}{1}$. Invert the new fraction to get $\frac{1}{4}$ and multiply: $3 \times 1 = 3$, and $4 \times 4 = 16$. **The answer is $\frac{3}{16}$.**

Now You Can Do It

$$6 \div \frac{2}{5}$$

$$\frac{2}{5} \div \frac{4}{5}$$

$$\frac{12}{25} \div \frac{3}{5}$$

$$\frac{4}{7} \div 5$$

$$3\frac{1}{8} \div \frac{2}{3}$$

$$\frac{2}{3} \div 6$$

WHAT ARE DECIMALS?

A special type of fraction is called a *decimal fraction.* That is any fraction that has 10, 100, 1,000, 10,000, etc., as a denominator.

$$\frac{1}{10} \quad \frac{1}{100} \quad \frac{1}{1{,}000} \quad \frac{1}{10{,}000}$$

$$\frac{2}{10} \quad \frac{2}{100} \quad \frac{2}{1{,}000} \quad \frac{2}{10{,}000}$$

$$\frac{7}{10} \quad \frac{7}{100} \quad \frac{7}{1{,}000} \quad \frac{7}{10{,}000}$$

Like other fractions, decimal fractions can be expressed by using a decimal point, which is a period that appears to the left of the number that represents the fraction. We don't write commas in decimal fractions.

$$\frac{1}{10} = .1$$

$$\frac{1}{100} = .01$$

$$\frac{1}{1{,}000} = .001$$

$$\frac{1}{10{,}000} = .0001$$

$$\frac{1}{100{,}000} = .00001$$

A mixed number with a decimal fraction can be written like this:

$$1\tfrac{3}{10}$$

Or like this:

$$1.3$$

The number of figures to the right of the decimal point tells us whether we're dealing with tenths, hundredths, thousandths, or whatever. One figure means the decimal fraction stands for tenths. Two figures, and it means hundredths. For example:

.3 = three tenths
.34 = 34 hundredths
.345 = 345 thousandths
.3456 = 3,456 ten thousandths
.34567 = 34,567 hundred thousandths

Okay. What if you want to write three hundredths? Use a 0 to mark the tenths place, just to the right of the decimal point:

.03 = three hundredths
.003 = three thousandths
.0003 = three ten thousandths
and so on to infinity. . . .

What if a zero appears in the far right-hand column of a decimal number? It may change the way you read the number, but it doesn't change the *value* of the number.

.5 = five tenths = $\frac{5}{10}$
.50 = fifty hundredths = $\frac{50}{100}$
.500 = five hundred thousandths = $\frac{500}{1,000}$

By canceling 0s, you can see that all these numbers have the same value:

$$\frac{500}{1000} = \frac{50}{100} = \frac{5}{10}$$

$$.500 = .50 = .5$$

This is the reason that your calculator may surprise you when you add \$2.25 and \$.25—it gives you an answer of 2.5. Where did the fifty cents go? Nowhere: to the calculator, .5 is the same as .50, and unless you tell it otherwise, it doesn't know you're thinking in terms of "cents."

Now You Can Do It

Write these numbers in numerals:

four tenths
fifteen hundredths
twenty-seven thousandths
eighty hundredths
one hundred sixty-four ten-thousandths
forty-seven millionths

Write a two-digit number, a three-digit number, four-digit number, and a five-digit number that all have the same value.

ADDING AND SUBTRACTING DECIMAL NUMBERS

Adding and subtracting decimals is just like adding and subtracting ordinary numbers. The only difference is that, in addition to keeping track of your place values, you also pay attention to where the decimal point falls. In other words, you keep track of place values *on either side of the decimal point.* To do this, you simply write a column of numbers to be added or subtracted so that the decimal points are aligned one below the other.

We can start with 2.61 plus 3.21 plus 24.12.

2.61 3.21 24.12 **29.94**	THINK: Write the numbers so their decimal points line up. This will allow you to keep the place values straight. Now add in the usual way, placing the decimal point right beneath the ones in the problem.

Try adding 3.87, 7245.789, 42.6, and 256.21.

3.87 7245.789 43.6 256.21 **7,549.469**	THINK: Keep the decimal points straight. Picture the blanks to the right of the decimal numbers as 0s: 3.87 = 3.87(0). Now you can add these the Human Calculator's way, from left to right. You quickly see that **the answer** is **7,549.469.**

Subtracting is just as simple. Try 8.69 minus 4.23:

8.69 −4.23 **4.46**	Subtract these as you would any number. Place the decimal point in the answer under the decimal in the columns above it.

If the decimal fraction in the subtrahend is longer than the decimal in the minuend, then add 0s to fill in the blank spaces:

$$\begin{array}{r} 8.69 \\ -4.237 \\ \hline \end{array} \quad = \quad \begin{array}{r} 8.690 \\ -4.237 \\ \hline \mathbf{4.453} \end{array}$$

$$\begin{array}{r} 43.86 \\ -7.62476 \\ \hline \end{array} \quad = \quad \begin{array}{r} 43.86000 \\ -\ \ 7.62476 \\ \hline \mathbf{36.23524} \end{array}$$

Remember: Keep the decimal points in line and the decimal numbers in the correct columns below each other. Visualize!

Now You Can Do It

Find the sum of 4.8, 27.635, 234.21, and 867.96.

Add 87.6 plus 4.325 plus 18 plus 102.33.

Find the difference between 876.8255 and 309.3346.

Subtract 98.254 from 149.15.

Add 34.2 to the difference between 45.9 and 67.

MULTIPLYING DECIMAL NUMBERS

Use any strategy that works for you in multiplying ordinary numbers. Then figure out where the decimal point goes in the answer by counting up the number of total number of digits to the right of the decimal points in the multiplier and the multiplicand. This is the number of digits that go to the right of the decimal point in the answer. An easy starting point is 2.4 × .12.

2.4 ×.12 **.288**	Start on the right with 2 × 4 = 8. Then cross-multiply and add the products.
2 × 2 = 4 4 × 1 = 4 8	Place this 8 in the answer as the middle digit. Multiply the left-hand digits to find the answer's left digit.
2 × 1 = 2	
.288	There are three digits to the left of the decimal points in the problem: 4, 1, and 2. So **the answer** has three digits to the left of its decimal point: **.288**

DIVIDING DECIMAL NUMBERS

There are two ways to do this. In either strategy, you divide the numbers just as you would whole numbers.

With method number one, count the digits to the right of the decimal points in the divisor and the dividend. Then subtract the divisor's number from the dividend's. This gives you the number of digits to the right of the decimal point in the answer.

.9)1.08	**THINK:** 108 ÷ 9 = 12.
↑ ↑↑	Count the digits in the decimals (3).
2 −1 1 **1.2**	Subtract the number of digits in the divisor from the number of digits in the dividend. Place the decimal point that number of places (1) from the right in the answer.

To use the second method, write the division problem in the conventional way:

$.9\overline{)1.08}$

Now, with a pencil or with your mind's eye, move the divisor's decimal point all the way to the right. Then move the dividend's decimal point the same number of places to the right.

$.9.\overline{)1.0.8}$

Divide as usual, keeping the place values straight:

$$\begin{array}{r} 1.2 \\ 9\overline{)10.8} \end{array}$$

Write the decimal point in the answer directly above the decimal point in the dividend, and you will have it in the right place.

Now You Can Do It

Multiplying and dividing decimals is just a matter of keeping track of place values. Watch your step with these:

47×86.3 $98.6 \times .376$ 54.11×86.22 9×3.857

$256 \div 82.7$ $97.5 \div .5$ $422.678 \div .82$ $.233 \div 3$

CONVERTING FRACTIONS TO DECIMALS

You can change *any* fraction into a decimal, simply by dividing the denominator into the numerator. For example:

$$\frac{4}{5} = 4 \div 5 = .8$$

How do you divide a larger number into a smaller number? Create a decimal point in the dividend and add 0s after it. The resulting quotient will have a decimal point, too. Like this:

$5\overline{)4} = 5\overline{)4.0}$

$$\begin{array}{r} .8 \\ 5\overline{)4.0} \end{array}$$

THINK: I can't divide 5 into 4. But I could divide 5 into 40. Now, 4 is the same as 4.0. If I add a decimal place, I *can* divide 5 into 4.0. The answer will have a decimal point in it.

If you wanted to find the decimal equivalent to 4/50, you would divide 4 by 50:

$$50\overline{)4} = 50\overline{)4.00}$$

$$\begin{array}{r} .08 \\ 50\overline{)4.00} \end{array}$$

As with multiplication, we can insert 0s just to the right of the decimal point to fill spaces.

This relationship between decimals and fractions greatly simplifies things when you have to calculate with strange-looking fractions. To multiply $5\frac{8}{15}$ times $4\frac{6}{7}$, for example, you can convert both sides to decimals and move forward:

$$5\frac{8}{15} = 5.533$$

$$4\frac{6}{7} = 4.85$$

$5.533 \times 4.85 = 26.83505$, or approximately 26.8

$$26.8 = 26\frac{8}{10} = \mathbf{26\frac{4}{5}}$$

This type of conversion is suspect.

As a matter of fact, you don't need a weird-looking fraction to take advantage of decimals. To the contrary, some of the most common fractions convert into easy-to-use decimals. As we saw in Chapter 9, 1/4 of anything is the same as .25. And:

1/2 = .5
1/3 = .33 (and a string of 3s extending into the distance)
2/3 = .66 (and a string of 6s)
3/4 = .75
1/5 = .2
2/5 = .4
3/5 = .6
4/5 = .8
1/8 = .125
1/12 = .08 (approximately)

Now You Can Do It

Express these numbers in decimals:

$$\frac{9}{10} \quad \frac{6}{8} \quad \frac{9}{12} \quad \frac{5}{27} \quad \frac{128}{240} \quad \frac{5}{8}$$

What common fraction could you use to say that the town water tank is seventy-five hundredths full?

How many tenths of the moon are showing when it's at the half-moon phase?

If we sliced the biggest pizza in the world into 1,000 pieces and then ate 125 pieces, what fraction of the pizza would be left?

14 Percentages

THE WORD "PERCENT" comes from the Latin phrase *per centum*—"by the hundred." Percentage is a variation on decimal numbers. It is a way of figuring parts of a whole by hundredths.

The commercial world uses percentages so widely that understanding how to use them is an essential part of any educated person's knowledge. Dollars and cents are really a form of percentage—there are 100 pennies to the dollar, and when we speak of a half-dollar, a quarter, or 27 cents, we are really talking about 50 percent, 25 percent, or 27 percent of a dollar, respectively. The interest that you pay on your house or car loan is computed in percentage, as are your income taxes and the tip you leave for the waiter.

When we say something like "Twenty percent of the crowd was American," we mean that 20 of every 100 people were Americans. If there were 200 people in the crowd, 40 of them would have been Americans. Of 1,000 people, 200 would have qualified as Americans. All these proportions are the same as $\frac{1}{5}$. Let's examine how this works.

HOW TO EXPRESS A FRACTION AS A PERCENTAGE

Any fraction can be converted into a percentage. If it's already a hundredth, you're in luck. Just express the hundredth as a decimal, delete the decimal point, and place the percent sign (%) after it.

$$\frac{14}{100} = .14 = \mathbf{14\%}$$

$$\frac{5}{100} = .05 = \mathbf{5\%}$$

$$\frac{1.25}{100} = .0125 = \mathbf{1.25\%}$$

To express other fractions as percents, first convert them into decimals. Then shift the decimal point two places to the right—as we did in the examples above.

$\frac{2}{5} = .4$

$.4 = \mathbf{40\%}$ Move the decimal point two places to the right.

$\frac{15}{16} = .93$ Move the decimal point.

$.93 = \mathbf{93\%}$

$\frac{14}{360} = .038$ Move the decimal point.

$.038 = \mathbf{3.8\%}$

$1\frac{3}{5} = \frac{8}{5}$ A percentage can be more than 100%.

$\frac{8}{5} = 1.6$ This happens when you start with a mixed number or an improper fraction.

$1.6 = \mathbf{160\%}$

Because "percent" means "per hundred," 1% means one in every hundred, or $\frac{1}{100}$.

$$10\% \text{ means } 10 \text{ in every } 100, \text{ or } \frac{10}{100}$$

$$25\% \text{ means } 25 \text{ in every } 100, \text{ or } \frac{25}{100}$$

$$98\% \text{ means } 98 \text{ in every } 100, \text{ or } \frac{98}{100}$$

If 100% represents the *whole* of some quantity or object, then 160% means you have more than the whole you started with. Let's go back to our giant pizza for an example.

We cut it into 100 slices, each piece about enough to feed one person. We thought that would be enough for the 100 people we invited to our party, but some guests brought their friends. So actually, 160 people showed up. We need 160 slices of pizza—or 160% of the pizza we started with.

Now You Can Do It

Practice converting fractions to percentages with these:

$$\frac{8}{10} \quad \frac{8}{9} \quad \frac{5}{20} \quad \frac{7}{8} \quad \frac{25}{40} \quad \frac{27}{60}$$

$$\frac{127}{60} \quad \frac{48}{95} \quad \frac{95}{45}$$

HOW TO EXPRESS A PERCENTAGE AS A DECIMAL FRACTION

Easy as pizza pie: just move the decimal point in the other direction. Instead of shifting it two places to the right, you move it two places to the left.

15(.0)% = .15
97% = .97
25% = .25
.6% = .006
3.33% = .0333
176.2% = 1.762

Now You Can Do It

What percentage is represented by these decimal fractions:

.48 .22 .50 .87 2.25

Hold this thought: we'll be using it soon. But first, consider one more point.

HOW TO EXPRESS A PERCENTAGE AS A COMMON FRACTION

To change a percentage into a common fraction, write the percentage number as a numerator over 100 as a denominator. Then reduce the resulting fraction to its lowest terms.

$$20\% = \frac{20}{100} = \frac{1}{5}$$

$$14\% = \frac{14}{100} = \frac{7}{50}$$

$$62.5\% = \frac{62.5}{100} = \frac{5}{8}$$

Now You Can Do It

What fractions are these percentages?

18%   33.3%   10%   90%   82%

CALCULATING WITH PERCENTAGES

You know that 47 percent of the students in your school plan to vote for you for student body president. If there are 850 students in the school, how many votes will you get?

To figure this out, you need to find 47 percent of 850. You know 47 percent is a fraction—it's the same as .47. So to find the size of your constituency, you simply multiply 850 times .47.

$$\begin{array}{r} 850 \\ \times\ .47 \\ \hline \mathbf{399.50} \end{array}$$

399.50 people plan to vote for you.

Round the answer off to 400 and you'll have an accurate count of your fan club.

So, to find a percentage of a number, change the percent to a decimal and multiply by the number.

Suppose the question goes the other way around. Suppose 170 students have said they will vote for you. You want to know what percentage of the total student body that is. How do you figure what percentage one number is of another?

Simple. Take the number that follows the question "What percent is . . ." (in this case, 170) and make it the numerator of a fraction. Make the other number the denominator. Reduce the fraction as far as possible, and then change it to a decimal number. All that's left is to move the decimal point two places to the right.

What percent is 170 of 850 students?

$$\frac{170}{850} = \frac{1}{5}$$

$$1 \div 5 = .20$$

$$.20 = 20\%$$

Some fractions are easily recognized as percentages. If you know that some quantity is $\frac{1}{4}$ of a whole, for example, you recognize that it's 25 percent of the whole. And vice versa—25 percent of something is a quarter of it.

Say your whale-watching club has spotted a pod of 48 whales off the coast of Seattle. Reporters say 25 percent of the pod has been spotted swimming up Puget Sound. How many whales were seen?

Knowing that 25 percent is the same as $\frac{1}{4}$, you quickly see that 48 ÷ 4, or 12 whales, have split off from the main group.

Here are percentages that you should learn to spot instantly as easy-to-work-with fractions:

$$10\% = \frac{1}{10}$$

$$20\% = \frac{1}{5}$$

$$25\% = \frac{1}{4}$$

$$33\frac{1}{3}\% \text{ (or, approximately, } 33\%) = \frac{1}{3}$$

$$40\% = \frac{2}{5}$$

$$50\% = \frac{1}{2}$$

$$66\frac{2}{3}\% \text{ (or, approximately, } 66\%) = \frac{2}{3}$$

$$75\% = \frac{3}{4}$$

$$80\% = \frac{4}{5}$$

So, if someone asks you to find 10% of a number, just divide the number by 10.

To find:

20% of a number, divide by 5.
25%, divide by 4.
$33\frac{1}{3}$%, divide by 3.
40%, divide by 5 and multiply by 2.
50%, divide by 2.
$66\frac{2}{3}$%, divide by 3 and multiply by 2.
75%, divide by 4 and multiply by 3.
80%, divide by 5 and multiply by 4.

Those are the really commonplace fractional equivalents to percentages. Everybody knows them—or ought to. But we human calculators know a few others. With these in mind, you can instantly come up with the answer to $37\frac{1}{2}$ percent of 72 (it's 27).

$$2\frac{1}{2}\% = \frac{1}{40}$$

$$12\frac{1}{2}\% = \frac{1}{8}$$

$$16\frac{2}{3}\% = \frac{1}{6}$$

$$37\frac{1}{2}\% = \frac{3}{8}$$

$$62\frac{1}{2}\% = \frac{5}{8}$$

What they mean is that to find:

$\frac{1}{2}$% of a number, you divide by 40.

$16\frac{2}{3}$%, divide by 6.

$37\frac{1}{2}$%, divide by 8 and multiply by 3.

$62\frac{1}{2}$%, divide by 8 and multiply by 5.

Now You Can Do It

What percent of 82 is 6?

How much is 20% of 320?

What percent of 150 is 25?

How much is 25% of 180?

What percent of 95 is $\frac{1}{3}$ of 95?

How much is $16\frac{2}{3}$% of 424?

MATH IN REAL LIFE

Percentage is one part of arithmetic that haunts your daily life. You'll find it everywhere you turn.

SALE! TAKE 40% OFF THE ALREADY DISCOUNTED PRICE!!!!

Boy, does that sound exciting! But what does it mean?

Suppose you go to the appliance store that is running this ad in the local paper. You find a refrigerator with a price tag of $688.99. (You realize, of course, that this is only 1 cent short of $689 and is so close to $700 that, by the time you pay a sales tax, you'll pay more than $700.)

But, the $688.99 price tag has been crossed out and marked down to $598.99. (Which, you recognize, is really about $600.)

According to the ad, though, if you buy this refrigerator, you get to take *another* 40 percent off the $598.99.

To figure out how much you would have to pay, you first take 40 percent of the round figure of $600. This is easy when you remember that 40 percent is the same as 2/5. All you have to do is divide $600 by 5 and then double the quotient ($600 ÷ 5 = $120; $120 × 2 = $240). An even simpler way to do this is to multiply by 40 and then divide by 100 (move the decimal point two places to the left): $600 × 40 = $24,000; $24,000 ÷ 100 = $240. This number—$240—is the 40 percent that will be taken off the original price. So, to find the new price, you simply subtract $240 from the old price of $600: $600 − $240 = $360.

So, you would pay $360 for a refrigerator whose original price was about $700. Not a bad deal.

Percentage also affects most Americans when they have to buy a car or a house. Few people have the resources to pay the amounts that these crucial items now cost, and so they have to buy them on time.

That means you take out a loan for the amount it costs to buy, say, a car. You have to repay the loan with a payment each month. But the bank that lends you the money earns its money by charging you something for the privilege of using its cash and paying it back over several months or years. This fee is called *interest.* The original amount you borrowed is called the *principal,* and the total of the interest and the principal is called the *amount.*

For example:	
You borrow $100.	**principal**
You agree to pay it back next month, with a fee of 10 percent for the privilege of using it.	**interest**
The amount you owe next month is $100 plus 10 percent of $100, or $100 plus $10, or $110.	**amount**

The percentage you pay in interest (in the example, 10 percent) is called the *rate.* Interest rates are usually expressed as a certain percentage of the principal for one year.

There are two kinds of interest: *simple interest* and *compound interest.*

Simple interest is the annual interest multiplied by the number of years the loan runs—that is, from the time you borrowed the money until the time you pay it back.

Suppose you borrow $150 at 8 percent interest, agreeing to repay it over two years. How much will you have to pay back, all told?

Well, in one year you will pay 8 percent of the principal. In two years, you will pay 2 times 8 percent, or 16 percent, of the original $150. That means you will owe 100 percent of the principal (the original $150) plus the 16 percent interest, or 116 percent of the principal. The amount you will owe will be 116 percent of $100, or 1.16 times 150, which is $174.00.

With *compound interest,* the principal draws interest for a specific period of time. At the end of this time, the principal and the accrued interest are added together, and this amount is taken as a new principal. The interest is figured on the sum for the next period of time. The interest is said to be *compounded.* If the period is one year, the interest is "compounded annually." If the period is six months, the interest is "compounded semiannually." For a period of three months, the interest is "compounded quarterly" (a quarter of a year). And so forth.

As you can imagine, the formula for figuring compound interest over a period of months or years is involved. Most people use a *table of compound interest* to figure amounts due on such loans. To use the one I have included at the end of this book, just read off the amount of one dollar at a specified rate for the period desired, and multiply that amount by the principal.

Let's imagine you want to open a lemonade stand. Unfortunately, you don't have the cash required to buy the frozen lemonade you need to get started. Your kid brother offers to lend you the money you need out of the untold hundreds he earned by selling all his Nintendo games in a garage sale.

You borrow $10 and agree to repay him in one month. He demands 5 percent interest per day, compounded daily. How much will you have to earn to pay him off?

Go to the 5 percent column. Run your eye down to the line for the period marked 30. The number you see in the 5 percent column is 4.32194. Multiply times 10 to get the amount owed: $43.22!

Now You Can Do It

You decide a loan in the amount of $43.22 will send you straight to the poorhouse. So you tell your kid brother that you will pay him 5 percent interest, all right—but 5 percent *simple* interest, figured over one month. If he goes for this, how much will you have to repay him on your $10 loan?

He rejects your offer. Instead, he proposes to charge you 5 percent per week compounded weekly. Now how much is he trying to extort from you? (Thirty days is 4.28 weeks, or slightly more than four periods.)

PART III

HANDY THINGS TO KNOW

15 Simplifying Math by Estimating

WE'VE SEEN A LOT of math in the past fourteen chapters, and we've learned how to figure things with some degree of precision. But now we have to look at two awful truths:

1. Sometimes it's just not necessary to get the answer right down to the nth decimal point.
2. Sometimes it's not *possible* to calculate an absolutely exact answer.

To start with that second hypothesis, consider the decimal fraction representing ⅓. We arrive at it by dividing 3 into 1. If you try that with long division, you can go on forever: .333333333 . . . and so forth. The decimal .33 is an approximation of ⅓. Similarly, the decimal version of 33 percent is .33333 etc. To figure a third of something, we either divide by 3 or multiply by .33. The answer is close enough for most applications.

Then we have numbers such as the square root of 2. You could calculate this figure until the sun burned into an ember, and you still would not have the precise value of $\sqrt{2}$. The distance around a circle is 3.14159 (etc.) times the diameter—but that number (known as *pi,* or π) cannot be calculated exactly.

Any answer is only as accurate as the figures you are given. If the original numbers are accurate only to a point, you waste your time figuring an answer to a precision beyond that point.

So, sometimes we are stuck with having to accept a reasonably close approximation of reality.

In other cases, it's actually useful or desirable to use an approximation. In the last chapter, for example, when we considered buying the cut-rate refrigerator, it was helpful for us to think of the original price as $700, rather than the $688.99 on the sales tag. And we had a better sense of what was real when we thought of the sale price as $600, rather than $598.99. These round numbers helped us see the real price—a hundred dollars more than the "$500" the mildly deceptive price was meant to make us think. Sometimes a good estimate is all you need, and it's enough to let you avoid a lot of needless figuring.

When we talked about adding from left to right, we noted that the most important digits appear in the leftmost columns, and that the value of the digits gets less significant as you move toward the right.

Take a number like 4,381—that's 4 thousands, 3 hundreds, 8 tens, and 1 unit. The digit 1 represents $\frac{1}{4{,}381}$ of the number. The 8 represents $\frac{8}{10}$ of 4,381; the 3 is $\frac{300}{4{,}381}$; and the 4 is $\frac{4{,}000}{4{,}381}$. If you rounded 4,381 to 4,380, the rounded-off number would be only $\frac{1}{4{,}380}$ off the correct figure. You could say it was *correct to three figures*. The fourth figure is not known to be correct, and so if you multiplied 4,380 × 2, the answer, 8,760, would also be correct only to three figures. But an error of $\frac{1}{4{,}381}$ is so small that in most cases you can accept it.

If you rounded 4,381 to 4,000, the approximated answer when you multiply by two would be 762 off plumb—that is, significantly changing the answer in the hundreds column. In many circumstances, this would be too inaccurate to be acceptable.

The "significant figures" for a calculation, then, depend on the circumstances. You must judge for yourself just how accurate an answer should be.

To identify the significant figures in a number, round it off as far as you need, starting with the right-hand digits. If the digit on the right is 5 or more, drop it (or change it to a 0) and add 1 to the digit to its left. This is called "rounding a number *up*." It means that whenever the right digit is 5, 6, 7, 8, or 9, you give the rounded figure a *higher* approximate value.

If the right-hand digit is less than 5, drop it without changing the next digit. This is rounding *down,* and it means that whenever the right digit is 4, 3, 2, or 1, you give the rounded figure a *lower* approximate value. Here's how it looks in action:

4,381
Rounded to the nearest ten: 4,380
Rounded to the nearest hundred: 4,400
Rounded to the nearest thousand: 4,000
6,627
Rounded to the nearest ten: 6,630
Rounded to the nearest hundred: 6,600
Rounded to the nearest thousand: 7,000

You see that you can keep moving toward the right, from the ones to the tens to the hundreds. By the time we've rounded 6,630 to 7,000, we're off the accurate figure by 370. That's quite a bit for some purposes—an engineer building a bridge could not accept this approximation, but it would do fine for a guess at the number of beans in a jar.

This principle works on decimal fractions, too. If the last figure on the right is less than 5, leave the digit to its left alone; if it's 5 or more, add to the next digit.

2.8653
Rounded to thousandths: 2.865
Rounded to hundredths: 2.87
Rounded to tenths: 2.9
Rounded to the nearest unit: 3

Now You Can Do It

If 23.6783 is rounded to 23.68, how far from accurate is the round number?

You need to know what fraction of an hour is represented by 38.67 minutes. What are the significant figures in 38.67?

Round:
34,894 to the nearest ten
12,555 to the nearest hundred

66.66 to the nearest unit
2,389,476.2338 to the nearest thousandth
79.7856 to the nearest unit
80 to the nearest hundred

Rounding numbers helps you precheck just about any calculation. If you have to multiply 1,496 × 98, for example, you see that the answer is not more than 1,500 × 100, or 150,000. If you come up with an answer that's more than that—or a lot less—you know you've made a mistake.

With division, you can eliminate a lot of work by omitting figures that do not affect the result's accuracy.

Let's look at a hefty example of long division:

```
               2,617.43
        ______________
   3,426)8,967,328.00
         6 852
         -----
         2 1153
         2 0556
         ------
           5972
           3426
           ----
           25468
           23982
           -----
            1486 0
            1370 4
            ------
             115 60
             102 78
             ------
              (etc.)
```

Suppose, however, that we did not need the answer correct to hundredths. Suppose instead we only needed an answer correct to four figures. Here's how we come up with a good round number for an answer and shorten the division process:

```
          2,617
3,426)8,967,328
        6852
        21153
        20556
          597
          342
          255
          238
        (stop)
```

Divide by 3,426 for the first couple of steps.

But here, stop bringing down numbers from the dividend, and drop the 6 from the divisor. 597 ÷ 342 = 1.

Subtract the product of 342 × 1 from 597 to get 255, and drop the 2 from the divisor. 255 ÷ 34 = 7. This is as far as you need to go.

Needless to say, this eliminates several steps. If the first digit in the decimal had been 5 or more, dividing by 34 would have produced an 8—neatly giving a rounded number correct to four figures.

Now You Can Do It

Find a sum, correct to three figures, for these numbers: 3,482; 1,765; 26; 814; 9,029.

Find the difference, correct to hundredths, between 1,278.97666 and 359.82963.

Find the product, correct to five places, of 347,896 and 27,433.

An ice rink measures 38.67 feet long by 50.28 feet wide. How many square feet does it contain, correct to 1⁄10 of a square foot?

Find an answer, correct to four figures, to 67,435 divided by 27.

Divide 100 by 3.14159, correct to .01.

ESTIMATION AS A SURVIVAL TECHNIQUE

You have to make an emergency run to the grocery store. Your family needs provisions for tonight's dinner and tomorrow's breakfast, but all you have in your pocket is a ten-dollar bill. You want to get past the checkout counter without being embarrassed.

In this case, the best thing to do is *always round up* instead of down. As you walk through the grocery store, round the prices of the things you need to the nearest quarter, fifty cents, or dollar, and keep a running total of the rounded prices in your head. Fudging the amounts on the high side means that your estimated grocery tab will be *more* than the actual cost, and so you should be safe at the checkout line. Let's try it:

Goodies	*Real Price*	*Rounded Price*	*Running Estimate*
Milk	$ .89	$1.00	$1.00
Cereal	2.25	2.50	3.50
Bacon	1.60	2.00	4.50
Lettuce	.79	1.00	5.50
Mayonnaise	1.17	1.25	6.75
Bread	.98	1.00	7.75
Margarine	1.14	1.25	9.00
Real Total:	**8.82**		

Your change is $1.18, almost enough to rent a movie tonight.

Another evening, you're hanging out in your favorite Greek restaurant with your buddies. You're feeling fairly rich—you do have more than $10.00 burning a hole in your pocket. Nevertheless, you'd like to have something left after tonight's frolic. Before you order, you make a ballpark estimate of how much this Dionysian feast is going to cost you. So you think about what you'd like for dinner, and again run up a mental adding-machine tape. Here, too, you fudge the amounts on the high side.

You know you can't get by without a hummus appetizer. This joint charges $2.50 for that, and they clip you another dollar for a basket of pita bread. You crave one of the incredible lamb dishes—the cheapest of those is $10.50, but you'd really like the awesome rack of lamb, $19.50. So, to the $3.50 you add $20, for $23.50. A half carafe of house wine is $4.90—add $5.00 to the $23.50 for a new total of $28.50. Baklava for dessert, $3.50, brings the bill close to $32.00, and you certainly need a cup of strong, sweet Greek coffee with which to swill down the baklava—that'll be $2.35, which, rounded to $2.50 brings the estimated total to $34.50. Now tips are on the bill *not including tax,* so you tip 15 percent on the rounded total of $35—that's 10 percent plus half of 10 percent, or $3.50 plus $1.75, which comes to a tip of $5.25. Since the bill *and tax* is around $36.50, round the $5.25 up to $5.50 and add for an even $42.00. Wow!

Let's see how the real tab would compare:

Hummus	$ 2.50
Pita bread	1.00
Rack of lamb	19.50
Wine	4.90
Baklava	3.50
Coffee	2.35
Subtotal:	33.75
Tip (15% of $35)	5.25
Tax (4.5%)	1.52
Total:	**40.52**

Your estimate was very close to reality. Good thing you inflated the prices for your guesstimate, or you might have found that your appetite was larger than your checkbook!

Now You Can Do It

You have a food budget of $50 per week for two people. Make a shopping list for a week's worth of groceries and go to your favorite supermarket. Walk through the aisles and seek out your items, keeping a running tab of what they might cost you. How close to $50 do you come? Do you have anything left over for a meal at a restaurant?

16 Quick-Check Your Answers

BEFORE WE BEGIN TO talk about how to check answers to addition, subtraction, multiplication, and division problems, I want you to know something extraordinary about the number 9.

Whenever a number is divided by 9, the remainder is equal to the sum of the number's digits. This strange fact, which can be proven mathematically, can be used in checking all four kinds of arithmetic computation.

Check it out:

$$151 \div 9 = 16, \text{ remainder } 7$$

$$1 + 5 + 1 = 7$$

The sum of the number's digits equals the remainder.

$$42 \div 9 = 4, \text{ remainder } 6$$

$$4 + 2 = 6$$

The sum of the number's digits equals the remainder.

"Oh yeah?" I can hear you saying. "Look here, Scott: 4,487

divided by 9 is 498, with a remainder of 5. But 4 plus 4 plus 8 plus 7 equals 23. That's a far cry from 5. How do you explain that?"

Well, there *is* an explanation. If the sum of the digits is more than 9, you have to cast out the nines from it, too. Add the digits, and then subtract 9 until the result comes to 9 or less.

$$4{,}487 \div 9 = 1{,}498, \text{ remainder } 5$$

$$4 + 4 + 8 + 7 = 23$$

23 does not equal 5

But: Cast out the 9s from 23.

$$23 - 9 = 14$$

$$14 - 9 = 5$$

The sum of the number's digits less all possible 9s equals the remainder.

Or, here's another way to deal with that: Cast out 9s by adding the digits in the sum of the digits:

$$4 + 4 + 8 + 7 = 23$$

$$2 + 3 = 5$$

The remainder of 4,487 ÷ 9 is 5.

$$5 = 5$$

"Okay, Scott. That sounds great. But what happens when the sum of the number's digits is evenly divisible by 9? Then the remainder is 0." Unless all the digits are 0s, it's not very likely that their sum will be 0!

If the sum of the digits is 9 or an even multiple of 9 (such as 18, 27, 36, etc.), then you count the remainder as 0.

45 ÷ 9 = 5, with no remainder

4 + 5 = 9

9 does not equal 0

But: Count 9 as a 0.

4 + 5 = 9, the same as 0

0 = 0

96,372 ÷ 9 = 10,708, with no remainder

9 + 6 + 3 + 7 + 2 = 27

27 is evenly divisible by 9 (27 ÷ 9 = 3)

Count 27 as a 0.

0 = 0

Or:

9 + 6 + 3 + 2 = 27

2 + 7 = 9

Count 9 as a 0.

96,372 ÷ 9 = 10,708, remainder 0

0 = 0

There's one more way to cast out 9s: by omitting them. As you add the number's digits, leave out the 9s and the combinations equal to 9 (such as 6 and 3, 4 and 5, 8 and 1, etc.)

192 ÷ 9 = 21, remainder 3

192: omit the 9; 1 + 2 = 3

3 = 3

6,367 ÷ 9 = 707, remainder 4

6,367: omit 6 + 3; 6 + 7 = 13; 13 − 9 = 4

4 = 4

Now, remember this technique. It forms part of the next four strategies.

CHECKING ADDITION

Here are five ways to check addition. Use the one that's easiest for you, under any given circumstances.

When you are adding only two numbers, the simplest way to check your addition is to subtract either of the numbers from the sum. If the remainder of your subtraction equals the other number in the addition problem, then your sum is correct. Let's try it on something simple:

$$\begin{array}{r} 10 \\ +\ 7 \\ \hline \mathbf{17} \end{array} \qquad \begin{array}{r} 17 \\ -\ 7 \\ \hline 10 \end{array} \qquad \begin{array}{r} 17 \\ -10 \\ \hline 7 \end{array}$$

Yes, the correct answer to 10 + 7 is **17**. What if, in a moment of inattention, we had written down 19 as the answer to 10 + 7?

$$\begin{array}{r} 10 \\ +\ 7 \\ \hline 19(?) \end{array} \qquad \begin{array}{r} 19 \\ -\ 7 \\ \hline 12 \end{array}$$

Uh oh. We see right away that 12 does not equal either of the numbers in our original problem, and immediately we know we made a mistake.

Let's try it on some larger numbers:

$$\begin{array}{r} 8{,}724 \\ +2{,}951 \\ \hline 11{,}675 \end{array} \qquad \begin{array}{r} 11{,}675 \\ -\ 2{,}951 \\ \hline 8{,}724 \end{array}$$

When you have more than two numbers in a problem, the most common way of checking addition is simply by adding the numbers in a different order. Add

$$\begin{array}{r} 2 \\ 8 \\ 3 \\ 4 \\ \underline{7} \end{array}$$

Of course, as a human calculator you spotted the numbers that added to 10, counted the tens, and tossed in whatever numbers were left:

$$\begin{array}{rr} 2 + 8 = & 10 \\ 3 + 7 = & \underline{10} \\ & 20 \\ & \underline{+ \quad 4} \\ & \mathbf{24} \end{array}$$

To check, you'd just add in a different combination: 4 + 7 = 11 + 3 = 14 + 8 = 22 + 2 = **24**. The pitfall here is that if you think 14 + 8 is 23, you'll get a wrong check figure, which will make your head ache.

Because addition works no matter in what order you add the numbers, you can check your addition by adding the digits in the opposite direction from the way you added the first time. If you added from top to bottom, check by adding again, this time from bottom to top.

Another method for checking addition is *casting out 9s*. The idea is to find all the ways that 9 appears in a number and then to get rid of all the 9s, in one of the ways we discussed above. The result is called a *check figure*.

First find the check figure for each of the numbers in the addition problem. Then find the check figure for the sum you obtained for the problem.

When you add the figures in the problem, you get a sum. If you find the *check figures* for all the numbers in the problem, you will find that the *sum of the check figures* results in a number that is the same as *the check figure of the sum!* Try it on 267 + 342 = 609.

The check figure for 267 is 6. The check figure for 342 is 0 (9 minus 9 = 0). The check figure for 609 is 6. Add the check figures for 267 and 342: 6 + 0 = 6. Compare this with the answer's check figure: 6 = 6.

When you add all the check figures that correspond to each of the numbers in the original problem, their sum should equal the check figure of the answer you got to the problem. To visualize that, think:

267	(2 + 6 + 7, minus 9s)	=	Check figure
+ 342	(3 + 4 + 2, minus 9s)	=	Check figure
Check figure of 609		=	Sum of check figures

(Check figure of sum = sum of Check figures)

For me, the easiest way to cast out 9s when the sum of the digits is more than 9 is to add the digits in the sum of the digits. It is also possible to simply subtract 9 over and over, until you arrive at a number lower than 9. (Obviously, if the digits don't add up to 9, you don't have a 9 to cast out.)

In this method, we will count a sum of 9 as 9. Let's try it:

3,425	3 + 4 + 2 + 5	=	14;	1 + 4	=	5
5,917	5 + 9 + 1 + 7	=	22;	2 + 2	=	4
231	2 + 3 + 1	=				6
4,482	4 + 4 + 8 + 2	=	18;	1 + 8	=	9
14,055						24

Okay, now we have a sum for our real numbers (14,055) and a sum for our check numbers (24). The first thing we do with these two figures is *cast out the 9s* in the check figure of 24. The easy way to do that is to add the two digits in the check figure: 2 + 4 = 6.

If our answer to the original problem is correct, we should be able to add the digits in the answer, cast out 9s, and come up with *the same number as the answer that we got when we cast out nines from our check figure of 24.* Let's try that:

14,055:

$$1 + 4 + 0 + 5 + 5 = 15$$

To cast out 9s from 15, add the digits: $1 + 5 = 6$, the check figure of the sum for the original problem.

Hallelujah! It's the same as the number we got when we cast out 9s from the sum of the check figure for each number in the problem: $6 = 6$.

You can see the potential problem here. If, in adding the original numbers, you accidentally added 231 as 321, the check figure you would get in casting out 9s would be the same: $2 + 3 + 1 = 3 + 2 + 1$. This mistake would give you an answer to your problem of 14,145.

$$1 + 4 + 1 + 4 + 5 = 15; 1 + 5 = 6$$
$$6 = 6$$

Heaven help us! Our check would show this error to be correct.

To detect an error caused by reversing the order of two digits, you can check an addition problem by *casting out 11s.*

For each number in the problem, start with the units digit and add every second figure, moving to the left. Then go to the tens column and add the rest of the digits. Subtract the second sum from the first sum to get your *check number.* If the first sum is smaller, add 11 or a multiple of 11 until you can subtract. To cast out 11s in the number 83,275, for example, do this:

83275	Start with units and add alternate digits:	$5 + 2 + 8 = 15$
83275	Go to tens and add alternate digits:	$7 + 3 = \underline{10}$
	Subtract the two results	5
	Your check figure would be 5.	

To check an addition problem, add the check figures for each number in the problem; they should equal the check figure of the original problem's sum.

3,425	(5 + 4) − (2 + 3) = 9 − 5	= 4
5,917	(7 + 9) − (1 + 5) = 16 − 6	= 10
231	(1 + 2) − 3	= 0
4,482	(2 + 4) − (8 + 4) = (6 − 12; 6 is smaller	
14,055	than 12, so to make a number larger than 12,	
	add 11 to the 6; this gives you 17 − 12)	= 5
	Add these check figures:	19
	Cast out 11:	− 11
	The sum of the check figures, minus 11s, equals:	8

Now find the check figure for the sum that you obtained in the original problem, 14,055:

$$(5 + 0 + 1) - (5 + 4) = 6 - 9$$

Six is less than 9, so you add 11 to 6:

$$6 + 11 - 9, \text{ which is } 17 - 9 = 8$$

Compare this with the figure that you got when you added up the check figures and cast out 11s:

$$8 = 8$$

If we had accidentally reversed 231 to read 321, its check number would have been (1 + 3) − 2 = 4 − 2 = 2. The total of the check figures would have been 10, and since 10 does not equal 8, we would have known something was amiss.

Notice that when mathematicians put numbers and signs inside parentheses, it means those figures should be calculated *before* you do the calculation indicated by the signs outside the parentheses. That is, first add 5 + 4 and 2 + 3; then subtract the results.

Casting out 9s or casting out 11s does not provide proof positive that your answer is right. But if the sums of your check figures don't tally, you can be sure there's a mistake somewhere.

The last method of checking addition is by adding up the columns and then, offsetting figures to allow for tens, hundreds, thousands and such (just as you might do in multiplication), adding the subtotals. The grand total should equal the sum obtained by whatever other method you used.

261	adding vertical columns left to right:	1	6	2	9
123		3	2	1	9
415		5	1	4	7
799		9	9	7	**799**

6,274	left to right	4	7	2	6	13
8,543		3	4	5	8	19
5,686		6	8	6	5	13
20,503		13	19	13	19	19
						20,503

Now You Can Do It

Show your stuff as a human calculator on these posers:

8 + 2 + 8 + 9 + 16 =

44 + 5 + 25 + 13 + 12 =

836 + 3 + 14 + 968 =

8,267 + 2,222 + 5,636 + 9,762 =

92 + 870 + 1,934 + 4,256 + 8,982 + 345 + 7 =

842 + 963 + 147 + 246 + 51,015 + 4,812 =

Now check three of the sums by casting out 9s or 11s. Decide for yourself which is the easiest of the four methods to check the remaining three, and prove them.

CHECKING SUBTRACTION

This is ludicrously easy. Since subtraction is reverse addition, how do you check it? Right: by adding!

Add the remainder to the subtrahend, and (if your calculations are correct) you'll get the other number in the problem.

$$\begin{array}{r} 8 \\ -\ 3 \\ \hline 5 \end{array}$$

Add remainder and subtrahend: 5 + 3 = 8.

Compare the minuend: 8 = 8.

If you feel that you absolutely *must* keep subtracting, you can check a subtraction problem by subtracting the remainder from the minuend. And if you're right, the result will be the same as the subtrahend.

8	Subtract the remainder from the minuend:
− 3	8 − 5 = 3.
5	Compare the subtrahend: 3 = 3.

You also can check subtraction by casting out 9s or casting out 11s. To cast out 9s, remember the sum of the check numbers of the remainder and the subtrahend should equal the check number of the minuend.

86	Check number of 64 plus check number of 22
−22	should equal check number of 86.
64	

Check number of 86:	8 + 6 = 14; 1 + 4	= 5
Check number of 22:	2 + 2	= 4
Check number of 64:	6 + 4 = 10; 1 + 0	= 1

Add the check number (4) of the subtrahend (22) to the check number (1) of the remainder (64). The result should equal the check number (5) of the minuend (86).

Check number of 22:	4
+ Check number of 64:	1
	5

5 = the check number of 86, 5

Now You Can Do It

Which of these answers is wrong?

2,705 − 297 = 2,408

87,673 − 57,944 = 92,729

235,934 − 68,292 = 168,642

843,209 − 234,586 = 690,613

476,704,398 − 25,873,252 = 21,797,246

87 − 99 = −12

Checking subtraction by adding is so simple that I think casting out 9s is more trouble than it's worth. However, this strategy *is* the easiest way of checking multiplication.

CHECKING MULTIPLICATION

Find the check numbers of the multiplicand, the multiplier, and the product. Multiply the first two check numbers and cast out 9s; the result should equal the check number of the product.

21	2 + 1 = 3: check number of multiplicand
×16	1 + 6 = 7: check number of multiplier
336	3 + 3 + 6 = 12; 1 + 2 = 3: check number of product

Multiply the check number (3) of the multiplier (16) by the check number (7) of the multiplicand: (21).

$$3 \times 7 = 21$$

Now cast out 9s from this figure, 21. You can do this by subtracting 9s until there you arrive at a number less than 9, or simply by adding the digits of the figure 21.

$21 - 9 = 12$	
$12 - 9 = 3$	This is one way to cast out 9s.
$2 + 1 = 3$	The other way is to add the figure's digits.

After casting out 9s, the check number that you get when you multiply the check numbers of the original numbers in the problem (the multiplier and the multiplicand's check numbers) is 3. Now compare that check number with the check number you derived from your original product, 336. Its check number, you will recall, was also 3.

$3 = 3$ Compare check numbers.

As with addition, you could accidentally reverse the digits in one of a problem's figures. Here again, the cure is to cast out 11s. Let's cast out 11s to get check numbers for these figures:

212	$(2 + 2) - 1 = 4 - 1 = 3$
$\times$ 163	$(3 + 1) - 6 = 4 - 6$ or
	$(4 + 11) - 6 = 15 - 6 = 9$
34,556	$(6 + 5 + 3) - (5 + 4) = 14 - 9 = 5$
$3 \times 9 = 27$	Multiply check numbers of multiplier and multiplicand.
$27 - 11 = 16$	Cast out 11s.
$16 - 11 = 5$	
$5 = 5$	Compare check numbers.

This strategy is easier and lots more fun than the traditional way of checking multiplication, because you get to use new numbers. Yes, you guessed it: The usual way is to exchange the multiplier and multiplicand and start all over again. That is, 212 × 163 should equal the same as 163 × 212. Tedious, but it works.

```
   212      163
 × 163    × 212
 -----    -----
   636      326
 1262      163
212       326
------   ------
34,556   34,556
```

It doesn't matter whether you use the conventional method (as we have here, just to show how much boredom awaits the unenlightened) or the faster strategies of cross-multiplication or complementary multiplication, you still end up repeating yourself. Casting out 9s or 11s gives you some variety and, once you get the hang of it, is very easy. Certainly for large numbers it's much quicker than remultiplying figures upwards of the hundred thousands.

The fourth and perhaps most obvious way to check multiplication is by reversing the process and dividing. If 3 × 9 = 27, then 27 divided by 9 = 3, or 27 divided by 3 = 9. When you divide the product by the multiplier, you should get the multiplicand. When you divide the product by the multiplicand, you should get the multiplier.

Now You Can Do It

Are any of these statements correct? What makes you think so?

9 × 276 = 2,503

27 × 23 = 621

8.2 × 4.4 = 3.608

333 × 666 = 222,778

1,789 × 9,871 = 17,650,219

42,368 × 965 = 40,664,120

878,943 × 722,654 = 635,171,684,722.

If you think two are right, you missed a trick question. Look harder at those numbers.

CHECKING DIVISION

I recommend that before you divide you try to *estimate the answer.* You'll be surprised at how many *big* mistakes this helps you to avoid.

Divide 897 into 54,793, for example. If you round off both numbers—900 into 55,000—you can arrive at a ballpark figure for the answer: Cancel the 0s and divide 550 by 9. This is a little over 61, and so the answer to the original problem is probably something near 61. The Human Calculator sees the answer is actually **61.08**.

Be sure, when you round off the divisor and the dividend, that you go in the same direction: Either increase both numbers or decrease both numbers. Don't increase one and decrease the other!

You can *keep tabs on long division as you go,* by checking the subtraction step by step. I do this by adding the partial remainders to the number above it:

11,859, r. 4

8)94,876

8

14	1 + 8 = 9; okay, bring down the 4.
8	
68	6 + 8 = 14; bring down the 8.
64	
47	4 + 64 = 68; bring down the 7.
40	
76	7 + 40 = 47; bring down the 6.
72	
4	4 + 2 = 6

Once you're finished, you can check your answer by multiplying it by the divisor. The result should equal the dividend.

47 16)768	Multiply 16 × 47 to get 752. This is different from the dividend. The alleged answer is wrong! Divide again and come up with 48 for an answer.
48 16)768	16 × **48** = 768; the **new answer** is correct.

Or, check the answer by dividing it into the dividend.

48 16)768	We might as well use an example we know is right.
16 48)768 16 = 16	Divide the dividend by the answer. If the result is the same as the divisor, you're on target.

Or you can use any of the human calculator strategies to double-check division. For example, dividing and dividing again by factors of the divisor:

$16 = 2 \times 8$	Break the divisor into factors.
$2\overline{)768}$ = **384**	Divide the first of them.
$8\overline{)384}$ = **48**	Divide by the second to confirm the answer.
$48 = 48$	

The easiest and fastest way to check division, especially if it involves large numbers, is by casting out 9s or 11s. Once the problem is solved and you think you have the answer, find:

the check number of the divisor
the check number of the dividend
the check number of the whole number in the quotient
the check number of the remainder

Then remember: The dividend's check number = (divisor's check number × quotient's check number) + remainder's check number.

Let's make up some larger numbers to experiment with casting out 9s:

$23\overline{)1{,}061}$ = **46, r. 3**	The **answer** is **46, r. 3**—I think.
$2 + 3 = 5$	Check number of divisor.
$1 + 0 + 6 + 1 = 8$	Check number of dividend.
$4 + 6 = 10; 1 + 0 = 1$	Check number of quotient.
$3 = 3$	Check number of remainder.
$5 \times 1 = 5; 5 + 3 = 8$	Multiply divisor's × quotient's check numbers and add remainder's check number.
$8 = 8$	Compare with dividend's check number. The answer is correct.

Now You Can Do It

Check these answers, using the method that works easiest for you:

$988 \div 19 = 52$

$2{,}993 \div 48 = 62$, r. 35

$7{,}200 \div 17 = 42.352$

$52{,}100 \div 99 = 526$, r. 26

$782 \div 23 = 34$, r. 5

$6{,}360 \div 76 = 83$

17 How We Measure Things

AMERICANS IN THE UNITED STATES grow up using the English system of weights and measures—ounces, pounds, feet, miles, and the like. We also speak of temperatures in degrees Fahrenheit. The rest of the world, including Canadians and Latin Americans, uses the metric system for weights and distances and speaks of temperatures in degrees Celsius. Although efforts to change our national habits have so far not taken hold, the United States approved the metric system in 1866, and metric is the official system used by scientific and technical departments of the United States government.

Because it is the international system of weights and measures, Americans need to understand metrics. Each of us will use it someday, whether to repair a foreign car, to cook up a French recipe, or to study the mysteries of outer space.

USING THE METRIC SYSTEM

Human calculators love the metric system, because instead of two or three sets of measures for volume, weight, and length, there's just one. Everything is expressed in decimal units, so to convert from one unit to another, you have only to move the decimal point.

All measures in the metric system are based on a single unit, the *meter*. This is supposed to represent one ten-millionth of the distance from the earth's equator to either pole. A meter is about 39⅖ inches long—a bit more than a yard.

Lengths shorter than a meter are measured in decimal fractions of the meter. The meter, then, is divided into ten parts, each called a *decimeter.* The decimeter is also divided into ten parts—each one one-hundredth of a meter—called *centimeters.* Centimeters are in turn divided into tenths—which are one one-thousandth of a meter—and that unit is called a *millimeter.*

Longer distances are measured in multiples of 10 meters. After the meter comes the *decameter*—10 meters. Next largest is the *hectometer* (10 decameters or 100 meters), and after that comes the *kilometer* (10 hectometers, or 100 decameters, or 1,000 meters). The kilometer is a little more than half a mile—actually, about ⅝ of a mile. The longest unit is called a *myriameter,* which is 10 kilometers.

These names come from Latin and Greek words meaning tenths, hundredths, thousandths and tens, hundreds, thousands. The same prefixes are used for measures of length, volume, and weight. When you speak of measures longer than the main unit, you're using terms of Greek origin; for measures smaller than the main unit, you're influenced by Latin. In a nutshell, they look like this:

milli- = thousandths (.001)
centi- = hundredths (.01)
deci- = tenths (.1)
Main Unit = one (1)
deca- = tens (10)
hecto- = hundreds (100)
kilo- = thousands (1,000)
myria- = ten thousands (10,000)

Most commonly used are thousandths (milli-), hundredths (centi-), units, and thousands (kilo-). Measurements in tenths, tens, hundreds, and ten thousands show up in science, but rarely appear in everyday use.

Length, then, is measured in multiples or fractions of meters: 1 meter (abbreviated m) equals

10 decimeters (dm)
100 centimeters (cm)
1,000 millimeters (mm)
.1 decameter (dam)

.01 hectometer (hm)
.001 kilometer (km)
.0001 myriameter (mym)

10 millimeters = 1 centimeter
10 centimeters = 1 decimeter
10 decimeters = 1 meter
10 meters = 1 decameter
10 decameters = 1 hectometer
10 hectometers = 1 kilometer
10 kilometers = 1 myriameter

Myriameters, hectometers, and decameters are rarely used—for most practical purposes, long distances are expressed in kilometers.

To figure out how many decimeters are in a meter, move the decimal point one place to the left. To see what part of a meter is represented by so many decimeters, move the decimal point one place to the right.

1 meter = 10 decimeters
1 decimeter = .1 meter
25 meters = 250 decimeters
25 decimeters = 2.5 meters
252 meters = 2,520 decimeters
252 decimeters = 25.2 meters

To change centimeters to meters, move the decimal point two places to the left. To change meters to centimeters, move the decimal point two places to the right.

1 meter = 100 centimeters
1 centimeter = .01 meter
25 meters = 2,500 centimeters
25 centimeters = .25 meter
252 meters = 25,200 centimeters
252 centimeters = 2.52 meters

To change millimeters to meters, move the decimal point three places to the left. And to change meters to millimeters, move it three places to the right.

1 meter = 1,000 millimeters
1 millimeter = .001 meter

25 meters = 25,000 millimeters
25 millimeters = .025 meter
252 meters = 252,000 millimeters
252 millimeters = .252 meter

To go in the other direction—toward larger numbers instead of smaller—simply reverse the movement of the decimal point. That is, to change meters to decameters, move the decimal point one place to the right. To change decameters to meters, move it one place to the left.

1 decameter = 10 meters
1 meter = .1 decameter
25 meters = 2.5 decameters
25 decameters = 250 meters
252 meters = 25.2 decameters
252 decameters = 2,520 meters

Similarly, to change meters to hectometers, move the decimal point two places to the right. To change hectometers to meters, move it two places to the left.

1 hectometer = 100 meters
1 meter = .01 hectometer
25 meters = .25 hectometer
25 hectometers = 2,500 meters
252 meters = 2.52 hectometers
252 hectometers = 25,200 meters

And to change meters to kilometers, move the decimal point three places to the left. To change kilometers to meters, move it three places to the right.

1 kilometer = 1,000 meters
1 meter = .001 kilometer
25 meters = .025 kilometer
25 kilometers = 25,000 meters
252 meters = .252 kilometer
252 kilometers = 252,000 meters

If you are an American driving in Latin America or Europe, you need to remember that a kilometer is *not* the same as a mile. In fact, it's a little more than half a mile, and when your speedometer reads "80 kph," you are holding up traffic on the Autobahn!

To measure area—such as the size of your front yard—we use *square meters.* One square meter is a meter long by a meter wide. It's also 10 decimeters long by 10 decimeters wide, or 100 square decimeters. Five square meters is 5 meters by 5 meters. One square decameter is 10 meters long by 10 meters wide, which is the same as 100 square meters. These measures also fit conveniently in a nutshell:

1 square meter (m^2)	=	100 square decimeters (dm^2)
	=	1,000 square centimeters (cm^2)
	=	10,000 square millimeters (mm^2)
1 square decameter (dam^2)	=	100 square meters
1 square hectometer (hm^2)	=	1,000 square meters
1 square kilometer (km^2)	=	10,000 square meters

One hundred square millimeters (mm^2) equal 1 square centimeter (cm^2).

A square decameter, in measuring land, is called an *are,* and the more commonly used square hectometer is called a *hectare.* One hectare is about 2½ acres.

Volume—the amount of space, for example, that is enclosed by a shoebox—is measured in cubic meters. Actually, a shoebox's volume would probably be measured in cubic decimeters or cubic centimeters.

1 cubic centimeter (cc or cm^3) = 1,000 cubic millimeters (mm^3)
1 cubic meter (m^3) = 1,000 cubic decimeters
10,000 cubic centimeters
100,000 cubic millimeters
1 cubic decameter (dam^3) = 1,000 cubic meters
1 cubic hectometer (hm^3) = 1,000 cubic decameters
1 cubic kilometer (km^3) = 1,000 cubic hectometers

Volume is different from *capacity.* Volume measures the amount of space inside a real or imagined object. Capacity measures the quantity of stuff a real or imagined space or vessel can hold. In the English system, volume is measured in cubic feet, cubic inches, cubic miles, etc. In the metric system, as we have seen, volume is a matter of cubic centimeters, cubic decimeters, cubic meters, etc. English capacity is measured in quarts, gallons, bushels, etc. In the metric system, capacity is measured in fractions or multiples of the *liter*.

A liter is a little less than one liquid quart. It's based on the cubic decimeter—one liter is an imaginary square container 1 decimeter high, 1 decimeter wide, and 1 decimeter deep. That is 1 dm × 1 dm × 1 dm, or 1 dm^3.

Like other metric measures, the liter is reckoned in decimal numbers, with the same prefixes: 1 liter equals

10 deciliters
100 centiliters
1,000 milliliters
.1 decaliter
.01 hectoliter
.001 kiloliter
.001 myrialiter

1 decaliter = 10 liters
1 hectoliter = 100 liters
1 kiloliter = 1,000 liters
1 myrialiter = 10,000 liters

Metric weight is measured in *grams*. The gram is based on the weight of one cubic centimeter of water. Now, water's weight changes according to its temperature, and so the temperature of the water scientists use to determine the weight of a gram is 4 degrees Celsius—at which water reaches its maximum weight. A gram is about $1/28$ of an ounce, and a kilogram is about 2⅕ pounds.

Fractions and multiples of the gram are designated in exactly the same way as other metric measures: 1 gram equals

10 decigrams
100 centigrams
1,000 milligrams
.1 decagram
.01 hectogram
.001 kilogram
.001 myriagram

1 decagram = 10 grams
1 hectogram = 100 grams

1 kilogram = 1,000 grams
1 myriagram = 10,000 grams

There is such a thing as a *metric ton,* which is the weight of one cubic meter of water at 4 degrees Celsius. A metric ton weighs about 2,205 pounds, a little more than the 2,000 pounds Americans commonly call a ton—which is more properly known as a *short ton.*

Speaking of degrees Celsius, the metric system also has a measure for temperature.

The two common temperature measures are *Fahrenheit*—most often used in the United States—and *Celsius* or *centigrade.*

In the Celsius system, the temperature at which water freezes is marked as 0° C., and the temperature at which it boils is 100° C. Temperatures higher and lower than the boiling point and freezing point of water are numbered accordingly. In Celsius degrees, the melting point of iron is at about 1,530° C., "room temperature" is about 20° C., and the temperature of the sun is about 6,000° C. A temperature below freezing is called a *negative temperature* and is marked by a minus sign (−). *Positive temperatures,* above freezing, are indicated by a plus sign (+). So you would say that the boiling point of water is +100° C., and the air turns liquid at −192° C.

The older Fahrenheit system has the boiling point of water at 212° F. and its freezing point at 32° F. There are 180 degrees between those two points. Zero degrees, Fahrenheit, is 32 degrees below freezing. On this scale, iron melts at 2,786° F., and the temperature of liquid air is about −459° F.

Now You Can Do It

How many meters theoretically cover the distance from the equator to the North Pole?

How many centimeters from the equator to the North Pole? How many decameters? Kilometers?

Convert 8,764,522.33 milligrams to grams. How many hectograms would this be?

CONVERTING METRIC TO ENGLISH MEASURES

If you go to Canada, you may see signs in gas stations advertising gasoline for $0.50. Sounds like a bargain—until you realize that's $0.50 *a liter*. Since a liter is about a quart, and there are four quarts in a gallon, the price is actually more than $2.00 a gallon!

You recover from the shock and buy enough gas to drive to the supermarket. There you find milk sold by the liter, cheese by the centigram, bulk rice by the kilogram, spices by the milligram, and pizza by the slice. What on earth is this stuff really costing you?

Puzzled, you retreat to your motel room and turn on the television. The weather announcer says tomorrow's temperature will be around 20° C. Should you wear a jacket or a sundress?

To put the cost of a liter, a gram, or a centigram into familiar terms, you need to know approximately what those metric measures mean in English measures. At the end of this book I've included two tables that summarize what you need to know about the English and metric systems, and a conversion table to help translate one to another. One of the most complete sets of conversion tables, by the way, is in the back of *Webster's New World Dictionary*.

To use a conversion table, look up *one* of the units you want to convert. See what that unit equals in the other system, and multiply by the other system's equivalent.

For example, suppose you need to know how many kilometers are in eight miles.

Look up 1 mile. You see that it equals 1.6 kilometers. If 1 mile = 1.6 kilometers, then 8 miles = 1.6 × 8 kilometers, or 12.8 kilometers.

What we've done, really, is set up a ratio and multiplied both sides by 8. This will work for any equivalent and any quantity.

1 cubic yard = .76 cubic meter
15 cubic yards = .76 × 15 = 11.4 cubic meters
1 cubic meter = 1.3 cubic yards
976 cubic meters = 1.3 × 976 = 1,268.8 cubic yards

As to what to wear on that trip to Canada, it's not difficult to convert Celsius to Fahrenheit, and vice versa.

One degree Celsius equals 9/5 degrees Fahrenheit, and 1 degree Fahrenheit is 5/9 degrees Celsius; and we know that 0° C. is 32° F. So:

To go from Celsius to Fahrenheit, multiply the degrees Celsius by 9/5 and then add 32.

To go from Fahrenheit to Celsius, *first* subtract 32 and then multiply the remainder by 5/9.

The Canadian weatherman said the temperature would be 20 degrees, and we're pretty sure he meant Celsius.

$$20 \times \frac{9}{5} = 36 + 32 = \textbf{68 degrees Fahrenheit}$$

Sounds like a nice day.

You were told the weather would be quite warm this week in the Canadian July, and so you brought clothes meant for 85° F. days. What would an 85-degree temperature read on the Celsius scale?

Start by subtracting 32:

$$85 - 32 = 53 \times \frac{5}{9} = \textbf{29.44 degrees Celsius.}$$

Now You Can Do It

Your kid brother has told your mom he has to stay home from school because he has a temperature of 37.4 degrees. You doubt he's *that* abnormal. Assuming he's giving her a Celsius figure, does he really have the body temperature of a cadaver? Or maybe he has a fever?

WHY BOTHER WITH METRIC?

Not only when you travel abroad do you need to know metric and Celsius measures. In the United States, the movement to adopt the metric system in all parts of daily commerce survives. Eventually, liters, grams, and meters will replace quarts, ounces, and yards. At that point, you as a consumer will have to be on the alert.

A liter is not exactly a quart. It's slightly more than a quart. And so if milk measured out in liter containers is sold for the price of a quart, you will get more milk. No big deal for one consumer—but over the years, and over hundreds, thousands, and millions of milk buyers, it will add up. Remember, even if you use round numbers for conversion: A yard is *not* a meter; a quart is *not* a liter; a ton is *not* a metric ton.

There are times when it is in your interest to be precise.

18 The Calendar Formula

HERE'S SOMETHING YOU WON'T find in just any math book. I can't resist including it because it's so much fun. Once you master this, you will become a confirmed human calculator.

With this strategy, you can ask a friend what day, month, and year he was born. Think for a few seconds, and then announce—with perfect accuracy!—which day of the week he came into the world.

Here's how it works.

First, you learn the *significant value* of each month in the year. These figures are as follows:

January	0
February	3
March	3
April	6
May	1
June	4
July	6
August	2
September	5
October	0

November 3
December 5

In addition, we assign each day of the week a number, starting with 0 for Sunday:

Sunday	0
Monday	1
Tuesday	2
Wednesday	3
Thursday	4
Friday	5
Saturday	6

Now, you take the year the person was born and drop the "19." If she was born in 1945, for example, take 45. Divide it by 4 and drop the remainder without rounding up.

$$45 \div 4 = 11 \text{ (drop the decimal remainder, .25)}$$

Add that back to the year:

$$11 + 45 = 56$$

To that, add the day of the month—our subject was born on May 7.

$$56 + 7 = 63$$

Now add the significant value for the month she was born in. The significant value for May is 1.

$$63 + 1 = 64$$

Divide this by 7, the number of the days in a week:

$$64 \div 7 = 9, \text{r. } 1$$

Now all you have to do is take the remainder and translate it into one of the assigned days of the week. Monday is 1, and so our friend was born on Monday, May 7, 1945.

I like to express this as a formula, which I call *the calendar formula.* In this statement, SV stands for "significant value."

$$\text{Day of the week} = \frac{(\text{year} \div 4) + \text{year} + \text{day} + \text{SV}}{7}$$

Once you have committed the significant values of the months to memory, this trick is incredibly easy. And incredibly amazing.

Before we leave this subject, there are a few details you need to know. When you're working with a leap year (that's any year that's evenly divisible by 4), you should subtract 1 from the significant value when you have to find a day in January or February.

The formula above works *only* for dates in the twentieth century (January 1, 1901–December 31, 2000). The reason is that the significant value for the months change as the centuries pass. The days rotate over a 400-year period. Every 400 years, a *leap century* occurs. Amazingly, every leap century begins with a Saturday. So any time you have a century that is divisible by 400, you know that January 1 of the first year in that century falls on Saturday. The year 2000 will begin on Saturday.

What if you want to use the calendar formula for a date that falls in another century? Well, with slight variations the formula works for any century that's reckoned by the Gregorian calendar. If you're looking for a day in the nineteenth century, add 2 to the significant value. To move forward a century from ours, subtract from the significant value: For the twenty-first century, you will subtract 1 from the SV.

Now You Can Do It

July 20, 1969: Neil Armstrong walked on the moon.

August 1, 1956: The Salk polio vaccine went into mass distribution.

December 8, 1914: Irving Berlin's first musical, *Watch Your Step,* opened on Broadway.

June 27, 1936: Franklin Delano Roosevelt said, with more prescience than he perhaps knew, "This generation of Americans has a rendezvous with destiny."

July 4, 1776: The Declaration of Independence was signed.

On which days of the week did these events occur?

More About How We Measure Years

Our 365-day year has not always been with us. Ancient Asians thought the year was 340 days long. They didn't know that the earth revolves around the sun, and so of course they couldn't measure a year in that way. Instead, they based their year on the passage of the seasons. They reckoned months according to the phases of the moon, and since the moon takes about 28 days to go from full all the way back to full, their months were also shorter than ours. They counted 12 months in a year, as we still do today.

The Babylonians, who developed a sophisticated knowledge of astronomy and mathematics, made closer observations. They figured a year was about 360 days, which they divided into 10 periods of 36 days each. Although they no longer corresponded to lunar phases, these intervals were still named after the moon.

By Roman times, several of the months had taken on names we use, in Anglicized form, to this day. Some of the months were named after Roman gods; others took on the Latin words for the numbers the Romans gave them.

January, from *Janus,* the two-faced god who looks forward and back at once
February, named after a Roman festival, *Februa*
March, after *Mars,* the Roman god of war

April, from the Latin *Aprilis,* perhaps "month of Venus"
May, probably from *Maia,* the goddess of increase
June, from the name of a prominent Roman family, *Junius*
July, the month of *Julius* Caesar
August, the month of *Augustus* Caesar
September, the Romans' seventh month (*septem,* "seven")
October, their eighth month (*octo,* "eight")
November, their ninth month (*novem*, "nine")
December, their tenth month (*decem*, "ten")

The Roman year began in March, and so October, November, and December were for them the eighth, ninth, and tenth months.

By Julius Caesar's time, people had learned that the year consisted of 365¼ days. Caesar decided to make the calendar conform to this. So he decreed that the year would henceforth consist of 365 days, that six hours in each year would be disregarded for three years, and that an entire day would be added to the second month in the fourth year to make up for the lost ¼ day. From this comes our *leap year*—the year that February contains 29 days.

Actually, the year contains 365 days, 5 hours, 48 minutes, and 46 seconds, a slight deviation from the Roman figure of 365 days, 6 hours. Over the centuries, this adds up. By the year 1582, the calendar was 10 days out of sync with the seasons. To fix this, Pope Gregory XIII decreed that 10 days would be erased from the calendar, and the day following October 4, 1582, would be called October 15. This brought the spring equinox—when daylight hours and nighttime hours are equal—to March 21. Most of the world's Catholic countries immediately adopted this Gregorian calendar.

The British, however, did not. They clung to the Julian calendar until 1752, by which time they were eleven days behind the Continent. Finally, Parliament agreed to strike eleven days from the calendar, and the day following September 2, 1752, became September 14. This is why documents written by English correspondents in the sixteenth, seventeenth, and eighteenth centuries do not agree in date with many of their continental contemporaries. In the Eastern Church, the old Julian calendar was retained until the twentieth century.

European nations number the centuries from the beginning of the Christian era. The years after the supposed year of Jesus Christ's birth are tagged A.D. (from the Latin *anno domini* "in the year of our Lord"). To be strictly correct, place the A.D. *before* the year:

"The American Revolution began in A.D. 1776." Years before Christ are marked B.C., which comes after the date: "Julius Caesar died in 44 B.C."

Technically, the last year of a century is the year ending in two 0s. Thus 1800 was not the first year of the nineteenth century but the last year of the eighteenth century. The twenty-first century begins in the year 2001. It will begin a new *millennium*—that is, a thousand-year period.

PART IV

MATH IN REAL LIFE

19 Tips, Tax, and Change

TIPPING

Tips are figured in round numbers on the before-tax tab: 15 percent give or take a few pennies. Precisely 15 percent of an \$8.75 check is \$1.31. Leave \$1.30, \$1.35, or—if the service was really *primo*—\$1.50.

Figuring .15 of any bill is easier if you first find 10 percent (.10) and then add 5 percent (.05). Since 5 percent is half of 10 percent, all you have to do is find 10 percent by moving the decimal point one place to the left, and then add half of that to the result.

Say the bill is \$14.62.

\$14.62 × .10 = \$1.46, or about \$1.45

½ of \$1.45 = \$0.73, or about \$0.75

\$1.45 + \$0.75 = **\$2.20, the tip**

Now You Can Do It

You fly into Denver for a business trip. The client is paying, so you take a taxi from the airport. The cabbie loads your three bags into the back of his car and drives you from Stapleton Airport to the Brown Palace Hotel. The fare is $14.95.

He and a bellboy load your bags onto a cart, and the bellboy wheels them into the lobby. You check in and are escorted by the bellboy to your room, where he unloads your bags and shows you where the bathroom is and how to operate the air-conditioning. You tip the bellboy 50¢ a bag.

You're starved, so you call room service and order a plate of cheese and crackers, to the tune of $10.50. A soft drink is another $2.25. Moments later, the food arrives at your door, delivered by a white-gloved waiter, who places the tray on a table and holds his hand out.

How much does your arrival in Denver cost your clients in tips? How could you save them money?

TAXES

Unlike tips, taxes are not optional. Many states and some cities charge a sales tax on restaurant meals, liquor, nonfood items purchased in grocery stores, and purchases such as clothing and appliances. In some places, such as New York City, sales taxes are high enough to significantly increase the cost of living there.

You figure the tax in much the same way as you do tips: on the net bill (before tipping, of course). To do this, you need to know how much tax your state levies, and, if there's a city tax, how much that is, too.

Suppose your state has a 4.5 percent tax, and your city a 1.2 percent tax. The easiest approach is to add them together: 4.5 + 1.2 = 5.7 percent. Convert this to a decimal by moving the decimal point two places to the left: 5.7 percent = .057.

To figure the tax precisely, multiply the total taxable bill by .057. To estimate the bill, round the multiplier up to .06.

Now, it's a rare municipality that charges tax on food items purchased in grocery stores. So, suppose your supermarket bill looks like this:

1 pound hamburger	$1.99
potato chips	1.28
canned soup	.67
canned soup	.67
canned soup	.67
fresh berries	.99
furniture polish	3.98
dishwasher detergent	4.26
plastic container	1.16
coffee	2.89
coffee maker	17.59
	$36.15

Your tax would *not* be 5.7 percent of $36.15. You would pay sales tax only on the nonfood items: furniture polish, dishwasher detergent, the plastic container, and the coffee maker. That would be $3.98 + $4.24 + $1.16 + $17.59, or $26.77. And 5.7 percent of that subtotal is $1.54. Add that to the subtotal of $36.15 to get your total grocery bill: $37.69.

There are many, many other kinds of taxes. Liquor and cigarettes, for example, carry various hidden taxes, depending on in what part of the country you purchase them—manufacturers, importers, distributors, and retailers are charged certain taxes that are passed along to you. State and federal income and social security taxes remove a chunk from your income.

COUNTING CHANGE

In some fast-food restaurants, the cash registers have keys marked with pictures of the items sold: hamburger, double hamburger, cheeseburger, salad, soft drink, french fries. Punch the milkshake key, and the machine adds $1.50 to the customer's tab. When you've punched in everything the customer ordered, the machine tells you what to charge him. He hands you a $20 bill to cover an

$11.45 meal. You punch in $20, and the machine tells you to give him $8.55 in change.

Very convenient. But, you know, these devices came into being not for your convenience, but because over the past twenty years business owners have found that the employees who staff their stores have a hard time making change. They blame this phenomenon on the sinking quality of American education.

It's time we brought a screeching halt to that attitude! If your mom and dad could figure correct change, so can you. And while we're not likely to make retailers replace their cash registers, you will know how to check the change you receive at stores and restaurants, and you will be able to count change correctly at your next garage sale.

To start, let's be sure you know how to count by 5s, 10s, and 20s. You certainly can count from 1 to 100 by 1s: 1, 2, 3, 4, 5, 6 . . . and so on. When you count by 5s, every other number ends in 0, and the rest end in 5:

5, 10, 15, 20, 25, 30, 35, 40, 45 . . .

Count by 10s, and every number ends in 0:

10, 20, 30, 40, 50, 60, 70, 80, 90, 100 . . .

Counting by 20s is the same as counting by 2s, except that you add a 0 to each number:

20, 40, 60, 80, 100 . . .

It seems obvious, but you would be amazed at the number of restaurant managers who have interviewed job applicants who can't do this.

When you make change, you start with the smallest units of change due and work up to the largest. The object is to hand the customer the largest pieces of money you can. For example, if you owe someone $8, you don't give her eight $1 bills; you give her three $1 bills and a $5 bill.

Making change is a simple matter of counting. You start your count with the amount charged and add up to the amount the customer gave you.

The customer's bill is $11.45. He gives you $20.00.

You say aloud, "That was $11.45."

Now you start taking money out of the bill. You start with a nickel. Say aloud, "Here's $11.50." Switch to quarters, and keep counting, "$11.75, $12.00." Now go to dollar bills and count till you come to a number ending in 5 or 0: "$13.00, $14.00, $15.00." Now you can fill in the rest with a $5 bill. As you hand it to him, you finish your count by saying, "And $20.00."

Easy, isn't it?

Let's try it again. You buy some items in a small-town drugstore whose proprietor still uses a cash register manufactured in 1946. The cost comes to $21.37. You give her two $20 bills, or a total of $40.00. The cashier counts your change:

"That was $21.37. Here's $21.38, 39, 40 (she puts three pennies in your hand); 50 (she gives you a dime), 75, $22 (she gives you two quarters); 23, 24, 25 (she gives you three dollar bills); $30 (she adds a $5 bill); and $40.00 (she finishes with a $10 bill)." If you count up your change, you will see it comes to $18.63, which is $40.00 less $21.37. She counted the change quicker than you could pull your calculator out of your pocket, and she was every bit as accurate as any electronic gadget.

Always count your change when you come away from a cash register. Machines and human beings make mistakes. You'll be surprised at how often incorrect change is offered—a little alertness will save you a lot of money.

Now You Can Do It

You go to dinner at a Greek restaurant and are charged $40.54. You have three $20 bills, which you hand to the waiter. He comes back with $19.66. Did he give you the right change? Figure it out by counting, rather than subtracting.

You are throwing a yard sale. To a single buyer, you sell one Nintendo game ($1.99), three jam jars and a peanut butter jar ($0.75 apiece), nine of the brand-new jeweled pig collars you bought awhile back ($2.98 apiece), and a stuffed elephant ($5.80). The customer gives you a $50 bill. Count his change for him.

To your amazement, the customer offers you a tip. If the generous customer gives you 15 percent of his total bill, how much is the tip?

20 Time: Jet Lag, Military Hours, Telling Time

LAST WEEK I TOOK a jet from Phoenix, Arizona, to San Diego, California. I left at 8:30 in the morning and arrived in San Diego at 8:35 the same morning. It was the shortest plane ride I've ever made!

Not really. In fact, the flight to southern California takes about an hour. But because San Diego is one time zone to the west of Phoenix, the local time there was an hour earlier than local Phoenix time. When I left Phoenix, it was 7:30 in California.

In the good old days, local time was *really* local. Time schedules were about as standardized as some of the early measures of weight and distance—some of which had to do with how far a man could holler across an empty plain. Each community kept its own time.

During the nineteenth century, people began to see the need for some standardized system of timekeeping. This was brought about by the invention of the steam engine and the subsequent development of long-distance, rapid railway routes. The patchwork of local times caused great confusion, not just for passengers but for shippers, train engineers, and railway operators. The problem was especially acute in the United States and Canada, where moving people and goods across the wide-open spaces meant crossing thousands of miles.

So, in 1884 representatives of twenty-seven nations met in Washington, D.C., where they agreed to adopt a uniform international time system. It's based on the relationship between *longitude* and time.

LONGITUDE AND LATITUDE

To understand this concept, let's look briefly at the way mapmakers locate positions on earth.

The earth is divided into two sets of standard reference lines. The lines are imaginary, of course—if you stand on the 45th parallel of latitude, you won't see a line running under your feet.

Latitude is a way of measuring distances north and south of the equator. It consists of parallel lines running in circles around the globe, starting at the Equator. These circles girdle the earth in layers, all the way up to the North Pole and all the way down to the South Pole. They are numbered in *degrees, minutes (indicated by the symbol ′),* and *seconds (″),* from 0° 0′ 0″ at the Equator to 90° at the Pole. If a person were standing halfway between the Equator and the North Pole, she would be at 45° north latitude. If she were halfway between the Equator and the South Pole, she would be at 45° south latitude.

You understand that a circle contains 360 degrees. If you draw a line around the globe that passes through each pole, you've drawn a circle. Now, if you travel along that line from the Equator to the North Pole, you've gone ¼ of the circle, or 90 degrees. Keep moving back down to the Equator, and you've gone ½ the circle, or 180 degrees. Continue down to the South Pole, and you will have traversed ¾ of the circle, or 270 degrees. And when you make it back to where you started, you will have traveled 360 degrees around the earth.

One degree contains 60 minutes (60′), and one minute contains 60 seconds (60″). These figures make it possible to locate a spot on the earth with some precision. The Tropic of Cancer, for example, is 23° 30′ north of the Equator. The border between Canada and the United States is at 49° north latitude. New York City's Museum of Natural History is at 40° 46′ 47.17″ north latitude.

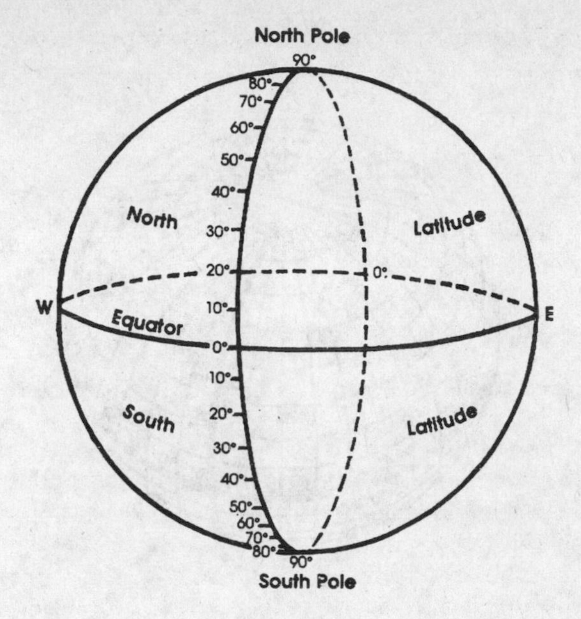

Longitude measures distances east and west. Longitude is pictured in circles, too, called *meridians*. Instead of being parallel, these circles all pass through the poles. Therefore, they draw closer together as they approach the poles, and spread wider apart closer to the Equator.

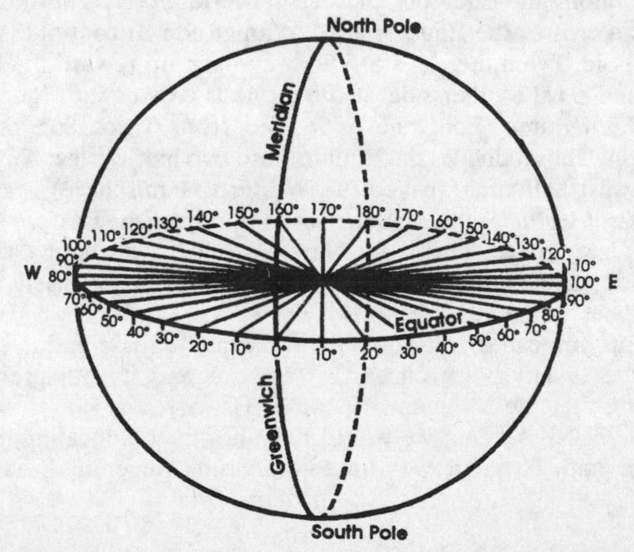

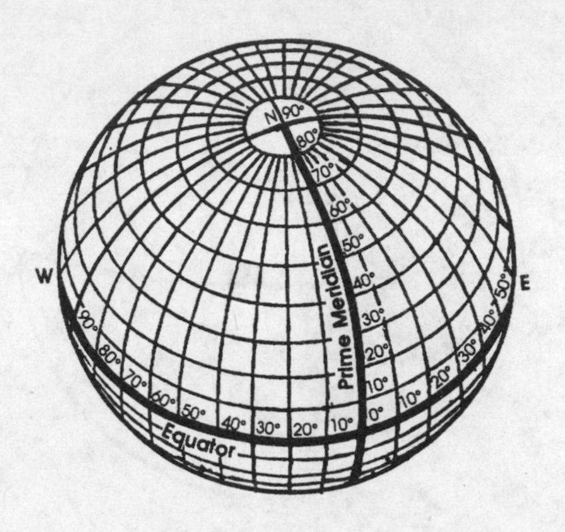

The starting point for measuring longitude is called the Prime Meridian. It passes through the British Royal Astronomical Observatory in Greenwich, England, and so it's sometimes called the Greenwich Prime Meridian. Degrees of longitude east and west are marked along the Equator. The Prime Meridian starts at the North Pole and crosses the Equator at 0° of longitude. It continues to the South Pole. From there it starts its way back up toward the North Pole, and on the other side of the globe it crosses the Equator at 180° of longitude. Longitude is figured from 0° to 180°, in each direction. Thus it divides the Equator into two half-circles. Anything in the west half-circle (called the Western Hemisphere) is said to be in *west longitude*. Places in the east half-circle (the Eastern Hemisphere) are in *east longitude*. New York City is about 74° west longitude, and the Museum of Natural History is precisely at 73° 58′ 41″ west longitude.

So, as you can see, if you have the exact latitude and the exact longitude of any place on earth, you can find it with precision. Knowing that the Museum of Natural History is at 40° 46′ 47.17″ N. and 73° 58′ 41″ W., we would have no trouble locating it on a detailed map. To find it, we would simply run a finger up the latitude scale to 40° 46′ 47.17″ north, and then go across to 73° 58′ 41″ west. The museum will be right where those two lines intersect.

. . . AND TIME AGAIN

Because the earth turns on its axis approximately once every twenty-four hours, longitude (lines drawn from the top to the bottom of the axis) has a great deal to do with time.

In one complete rotation, the earth turns through all 360 degrees of longitude. In one hour, it turns 360 degrees ÷ 24, or 15 degrees. So, one hour of time is equivalent to 15 degrees of longitude.

Picture the sun directly over the Greenwich meridian on the Equator. At that point, it's noon at 0° of longitude. But as we go east from the Prime Meridian, local time is one hour later for every 15 degrees of longitude. As we go west, local time is one hour earlier for every 15 degrees. Take a look at the figure below.

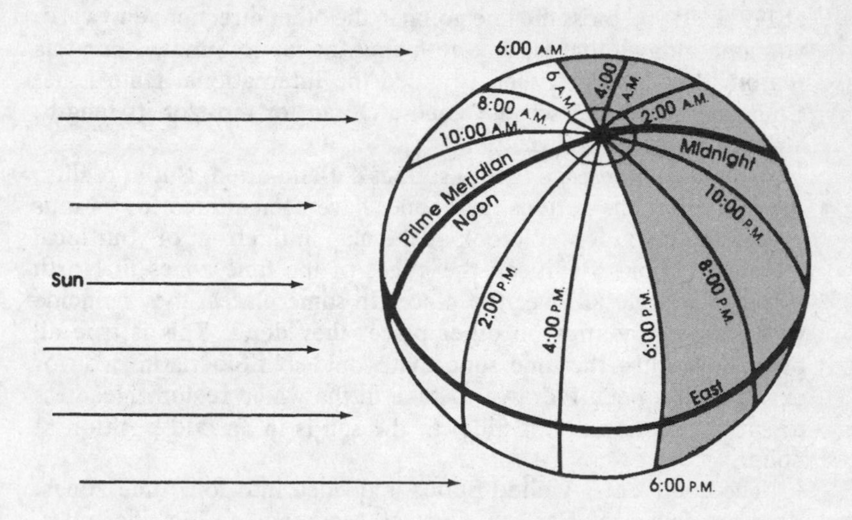

When the sun is directly over the Prime Meridian, it is noon at 0° of longitude and midnight on the other side of the globe, at 180°. At 90° west, the sun appears to be rising, and it's 6:00 in the morning. At 90° east, the sun is setting, and it's 6:00 in the evening. So you see that the time of day depends on *where you are* on the earth.

INTERNATIONAL TIME

The potentates who gathered in Washington back in 1884 to divide the earth into twenty-four *standard time zones,* started with the Greenwich Prime Meridian. Each time zone centers on a standard meridian of longitude, and these are set 15 degrees apart at the North Pole. So, each time zone represents one hour of time. If it's noon at Greenwich, it's 1:00 P.M. in the next time zone to the east. Two zones to the east, it's 2:00 P.M. Six zones east, it's 6:00 P.M. One time zone to the west, it's 11:00 A.M. Two zones to the west, it's 10:00 A.M. Six zones west, it's 6:00 A.M. As you go east from Greenwich, you add one hour of clock time for each time zone you pass through. As you go west, subtract one hour for each time zone.

All the way on the other side of the earth, where longitude measures 180° east and west, it's midnight at noon Greenwich time. And when you cross that line, you move into another day! If you're traveling west, you lose a day—it's Sunday at 179° W. and Monday at 179° E. If you cross the line going in the other direction, eastward, you gain a day—that is, you go from Monday to Sunday. For this reason, the 180th meridian is called the International Date Line. Conveniently, it crosses the Pacific Ocean for most of its length, and few people live near it.

Now, this time zone theory sounds cut-and-dried. But in reality, the borders of the various time zones have been altered for various special interests. If you'll look at the map in the front of your local telephone book, you'll see the edges of the time zones in North America wriggle all over the place. In some places, they coincide with state boundaries; in other places they don't. This is true all over the world—the time zone that would split Siberia in two, for example, has been redrawn to take in the whole region. Needless to say, in some parts of Siberia, the sun is in an odd position at noon.

The continental United States is divided into four time zones. They are called the Eastern, Central, Mountain, and Pacific zones. The easternmost provinces of Canada are in the Atlantic time zone. Alaska is one hour earlier and Hawaii is two hours earlier than Pacific time.

To figure the local time in another zone, count the number of zones east or west (not counting your zone) and add or subtract that many hours. When it's 4:00 P.M. in New York City, it's 3:00

P.M. in Des Moines, 2:00 P.M. in Denver, 1:00 P.M. in San Francisco, noon in Anchorage, and 11:00 A.M. in Honolulu.

TIME ZONES, JET PLANES, AND TELEPHONES

What does it all mean in practical terms?

Well, if we still got around by horse and buggy and communicated by smoke signals, it wouldn't mean much. But lately, we humans have taken to flying around the world at six hundred or seven hundred miles per hour. This means we can cross time zones in a matter of minutes. We talk to each other over telephones bouncing microwave signals off satellites—and that means I can call you at 10:00 P.M. my time and wake you up at 2:00 in the morning your time.

When I fly from New York to Paris, I cross six time zones traveling *east*. That means that when I arrive in Paris, I need to set my watch six hours *ahead* of New York time. If I fly from New York to Honolulu, I pass through five time zones traveling *west*. So, when I reach Hawaii, I set my watch five hours *behind* New York time. And if you're in Hawaii and I decide to telephone you from New York to let you know when I will arrive, I must remember not to call you at 9:00 A.M. my time—that would be 4:00 A.M. your time!

How will I know what time my plane will land in Paris, if it takes off from New York at 7:30 P.M.?

It takes about six hours to fly from New York City to Paris, and Paris is *east* of New York. So, I *add* six hours to 7:30 P.M.

7:30 P.M. + 6 hours = 1:30 A.M. New York time

Now, to adjust for Paris time, add the 6-hour time difference:

1:30 A.M. + 6 hours = **7:30 A.M. Paris time**

If I leave New York at 7:30 P.M. bound for Honolulu, what time will I land there?

First add the ten-hour flight time, then subtract the five-hour time difference.

Suppose it takes about 10 hours to fly from New York to Honolulu. I first add the flight time:

7:30 P.M. − 10 hours = 5:30 A.M. in the morning
New York time

Now, there's a five-hour time difference between Hawaii and New York. Hawaii is five hours *behind* New York. So, subtract five hours from New York time to get the local arrival time:

5:30 A.M. − 5 hours = **12:30 A.M. Honolulu time**

Some people's bodies get confused by these time changes. Boarding a plane at dawn and getting off a few minutes later—after six hours in the air—can make you feel tired, cranky, and out of sorts. Your body thinks it's bedtime, but the sun is telling you it's time to rise and shine. This is known as jet lag.

For most people who are bothered by jet lag, it feels worse to travel long distances east than to go west. As you can see by our Paris and Honolulu trips, when you go east—flying into the sun—you're more likely to arrive at a local hour much different from the hour that your internal clock is set to register. For this reason, if you have to fly from California to England, it's wise to stop for a day in New York to let your body catch up.

Many people like to leave their watches set at their back-home time. I find this counterproductive. The faster you can adjust psychologically to the local time, the quicker you will get back to normal.

MILITARY TIME

You and Granddad go out to the nearby Air Force base, where he is a member of the Air Force Reserve. On a bulletin board, you see a notice that the accounting office opens at 1400 hours.

Say what?

Actually, they mean that the office opens at 2:00 P.M. All branches of the United States military, as well as scientists around the world, use a twenty-four-hour scale for telling time. In this scheme, time is expressed in four-digit numbers, from 0100 (oh one hundred) hours to 2400 (twenty-four hundred) hours. Midnight occurs at 2400 hours, and the day begins at one minute after midnight, or 0001 hours. Noon is called 1200 hours. Instead of starting to count the hours over after noon, though, the military keeps counting toward 2400: 1:00 P.M. is 1300 hours; 2:00 P.M. is 1400 hours; 5:00 P.M. is 1700 hours.

You translate P.M. hours into military time simply by adding 12(00):

2:00 P.M. = 2 + 1200 = 1400 hours
9:00 P.M. = 9 + 1200 = 2100 hours

Minutes go in the tens and units places:

8:10 A.M. = 0810 hours
12:27 P.M. = 1227 hours
4:56 P.M. = 1656 hours
Quarter to two P.M. = 1345 hours
One minute after midnight = 0001 hours

NAUTICAL TIME

Mariners used to toll the time by the number of hours in a shift, or *watch.* Because someone has to watch a ship at sea twenty-four hours a day, crew members divided the day into four-hour watches. Each hour, the ship's clock tolls in multiples of two, up to eight bells. A shift would start at 12:00 noon or midnight. The first hour, 1:00, was marked by two bells. The clock chimes twice: *bong-bong.* At 2:00 A.M. and P.M., the clock chimes four times: *bong-bong bong-bong.* At 3:00, it chimes six times, and at 4:00, it chimes eight times, marking the end of a watch. The cycle starts over at 5:00, which is again two bells. Half hours are marked by a single chime. For example, 1:30 is three bells: *bong-bong bong.*

If you listened to the clock, you knew how long you had to work or rest, and how much time was left before the next meal:

1:00	two bells
1:30	three bells
2:00	four bells
2:30	five bells
3:00	six bells
3:30	seven bells
4:00	eight bells
4:30	one bell

5:00	two bells
5:30	three bells
6:00	four bells
6:30	five bells
7:00	six bells
7:30	seven bells
8:00	eight bells
8:30	one bell
9:00	two bells
9:30	three bells
10:00	four bells
10:30	five bells
11:00	six bells
11:30	seven bells
12:00	eight bells
12:30	one bell

This system seems quaint and even confusing to landlubbers, and probably not many modern sailors are familiar with it, either. But mariners who navigated by the sun and the stars were sensitive to the turn of the hours, and when "six bells" sounded, they knew instantly whether the time was 3:00, 7:00, or 11:00—A.M. or P.M.

Now You Can Do It

Granddad is going on a short assignment for the Air Force Reserve. He will leave his home in Seattle at 0945 hours and fly to San Antonio, Texas, arriving there at 1330 hours, local time. How long will he be in the air?

On a globe map, find a place located about 37° N. and 3°E. What city is located there, and what country is it in? When it is 1:00 P.M. where you live, what time is it there? This city is in the first time zone west of the Greenwich meridian.

How would an Army sergeant express "twenty to six in the evening" if she had to write it in an official report?

In what ways does an analog clock resemble the earth's longitude? Identify as many as you can.

Popeye the Sailor Man says dinner is at four bells. What time do we eat?

21 How Big *Is* the Back Forty?

"THE BACK FORTY" is probably what you call the yard in back of your house every time you have to mow it. That was what our agricultural forebears called the remote forty acres down on the back end of the farm. On a hot summer Saturday, that overgrown lawn seems as big as a forty-acre field!

You can figure out how many square feet the backyard really covers, and you can do it pretty easily even if your yard is a strange shape. Once you know how many square feet it contains, you'll know just how close to forty acres it really is.

Areas of flat shapes are measured in *square units*—square feet, square yards, square inches, square miles, square meters, square whatever. The sides of such a shape, called its *dimensions,* are given in *linear* measures—feet, inches, miles, meters, etc. In geometry, a flat shape is called a *plane figure;* it has only two dimensions, length and width. Objects with more than two dimensions are called *solid figures*.

Flat things, obviously, come in lots of different shapes. The way we figure the area of a plane figure depends on its shape.

Figures with three sides and three angles are called *triangles*. Four-sided, four-angled figures are called *quadrilaterals.* After that, Greek prefixes before the suffix *-gon* are used to say how many sides (and angles) the figure has. A *pentagon* has five sides (*penta-* means "five"); a *hexagon* has six sides; a *septagon* has seven; an *octagon* has eight; and so forth.

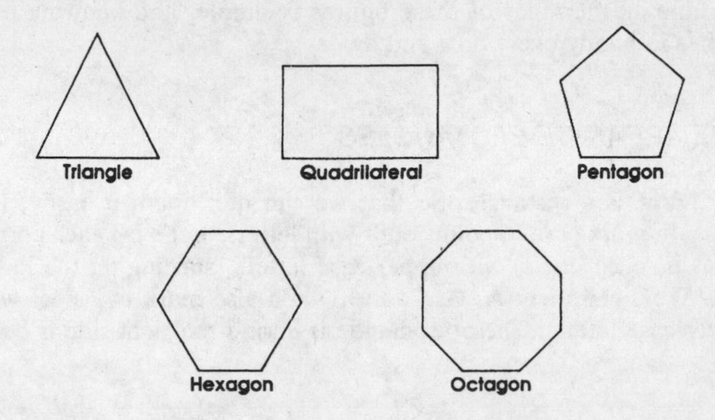

If all four angles of a quadrilateral are right (90-degree) angles, it is called a *rectangle*. And if all four sides of a rectangle are equal, it's a *square*. If the opposite sides of any quadrilateral are parallel, the figure is called a *parallelogram*. Rectangles and squares, for example, are parallelograms—but not all parallelograms are rectangles or squares. If only two sides of a quadrilateral are parallel and the other two go every which way, then the figure is called a *trapezoid*.

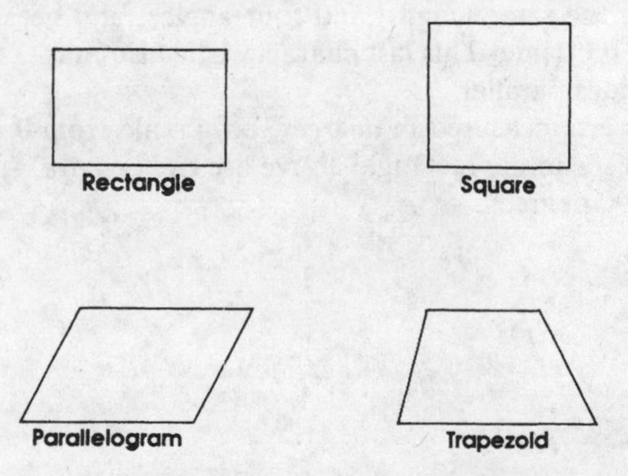

A perfectly round plane figure is a *circle*. Roundness is the only characteristic that distinguishes the circle, and so circles can differ only in size.

Finding the areas of these figures is simple, and knowing how comes in handy every now and again.

THE AREAS OF RECTANGLES

Below is a rectangle. So that we can talk about it easily, I'm going to mark certain points on it with letters. Let's tag each corner (that is, each *angle*) with upper-case letters, starting on the lower left. We'll call them A, B, C, and D. I'll also mark two sides with lower-case letters: The bottom side is *a,* and the right side is *b*.

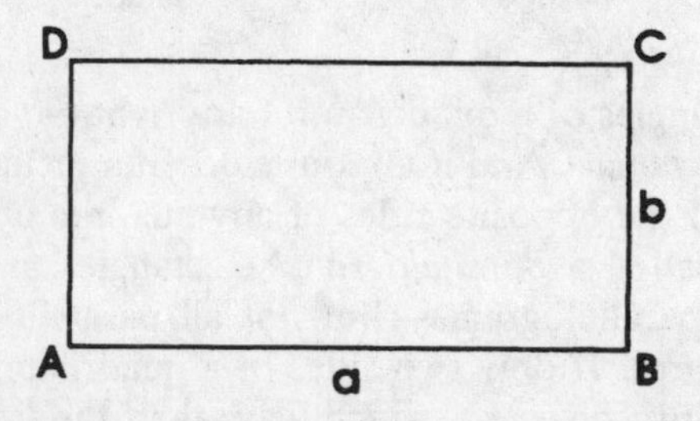

We know this thing is a rectangle because it has four sides (not necessarily the same length) and four angles, and because all four angles are the same. This last characteristic also makes each pair of opposite sides parallel.

Angles are measured in *degrees,* on a scale from 0 to 360. The angles you see in the rectangle above are *right angles.* A right angle measures 90 degrees.

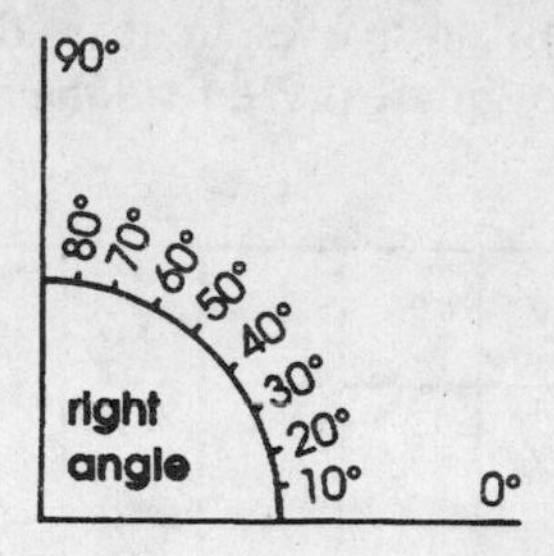

If our drawing were a square, the only difference would be that all four sides would be the same length. It would still have four 90-degree angles. As you can see, an oblong box is called a rectangle. A square is also a rectangle, but it is a rectangle whose four sides are equal.

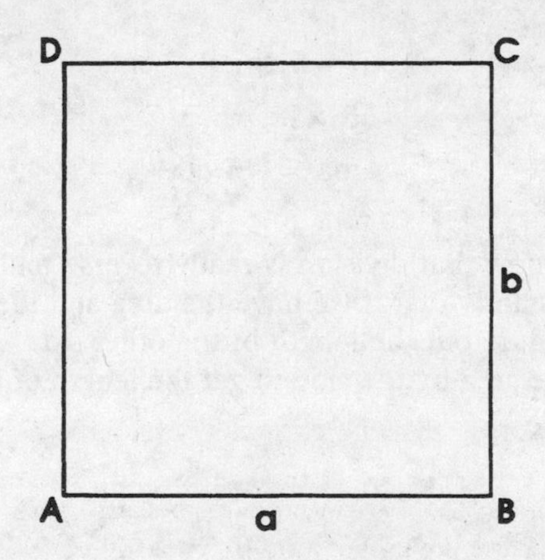

To find the area of a rectangle, just take the length of the base (from A to B, or, as mathematicians say, AB) and multiply it by the height (B to C, or BC). Because we've given AB the tag a and we've named BC b, we could simply say that the area = $a \times b$.

$$\text{Area} = a \times b$$

Okay. Suppose our rectangle is 5 centimeters long and 3 centimeters high. If we draw those even units inside the rectangle below, we see they amount to squares. We read the area as "square" units.

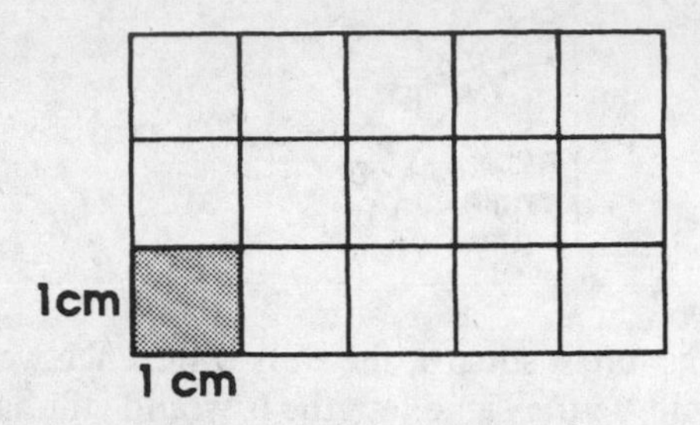

Multiply the length, 5, times the height, 3, to get the area in square units:

$$\text{Area} = 5 \text{ cm} \times 3 \text{ cm}$$
$$5 \times 3 = 15$$
$$\text{Area} = \mathbf{15\ cm^2}$$

Since we know that division is really reverse multiplication, we also realize instantly that if we have the area and the length of one side, we can figure out the length of the other side. All we have to do is divide the area by the side to get the length of the other side:

$$15 \text{ cm} \div 3 \text{ cm} = 5 \text{ cm}$$
$$15 \text{ cm} \div 5 \text{ cm} = 3 \text{ cm}$$

THE AREAS OF SQUARES

Amazingly simple, isn't it? Life is even easier if all four sides of your rectangle are equal, so that it is a square. Then a is the same as b, and to get the area of a square you—that's right—*square one of the sides.*

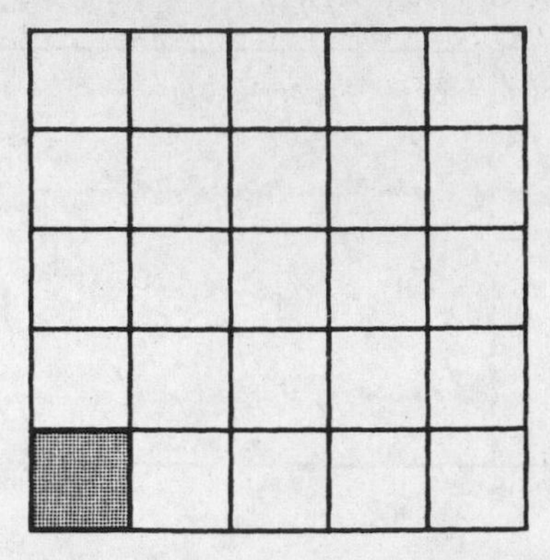

Say a rectangle is 5 inches long by 5 inches high. The area then is 5^2.

$$\text{Area} = a^2$$

$$\text{Area} = 5^2 = 5 \times 5 = \textbf{25 square inches}$$

If you know the area of a square and want to know how long its sides are, all you have to do is find the square root of the area.

$$a = \sqrt{\text{Area}}$$

So, given a square whose area is 64 square meters, you know that each side is 8 meters long. Why? Because the square root of 64 is 8.

THE AREA OF A PARALLELOGRAM

The figure below depicts a parallelogram.

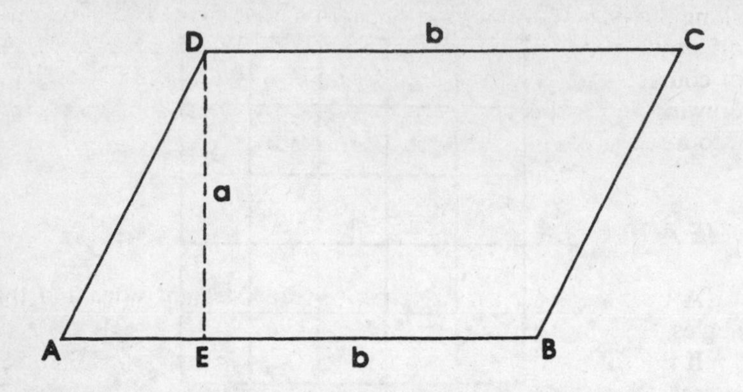

Once again, I've labeled the corners with upper-case letters. In this case, we *don't* find the area by multiplying one short side by one long side. Instead, we draw a perpendicular line from the top to the bottom sides, which is called the parallelogram's *altitude*. I'll mark it a. The bottom line is called the *base* of the parallelogram, and I'll call it b.

Find the area by multiplying the base times the altitude.

$$\text{Area} = \text{altitude} \times \text{base}$$
$$\text{Area} = a \times b$$

If the figure above has a base of 8 feet and an altitude of 5 feet, its area is what?

$$\text{Area} = a \times b$$
$$\text{Altitude} = 5 \text{ feet}$$
$$\text{Base} = 8 \text{ feet}$$
$$\text{Area} = 5 \text{ feet} \times 8 \text{ feet} = \mathbf{40\ square\ feet}$$

Why does this work? Notice that when you draw the altitude from the angle D, you form a triangle. If you cut that triangle off and moved it over to the right side of the figure—just sort of fit it along the line BC—you'd end up with a rectangle. The part you cut off would neatly fill up the space needed to form a rectangle. And of course, you'd find the area by multiplying $a \times b$. In effect, by drawing and measuring the altitude, you convert the parallelogram into a rectangle.

THE AREAS OF TRIANGLES

A triangle is a plane figure with three straight sides and three angles. They come in many shapes.

If the three sides are equal in length, the triangle is called *equilateral*. Sometimes equilateral triangles are called *equiangular*, because if all three sides are equal, all three angles will be equal, too. The figure below shows an equilateral triangle.

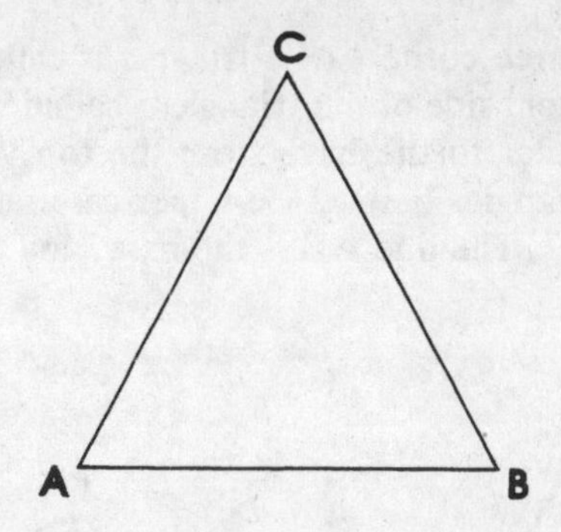

If the sides and angles are *not* equal and one of the angles is larger than a right angle, then the triangle is called an *obtuse* or *oblique* triangle.

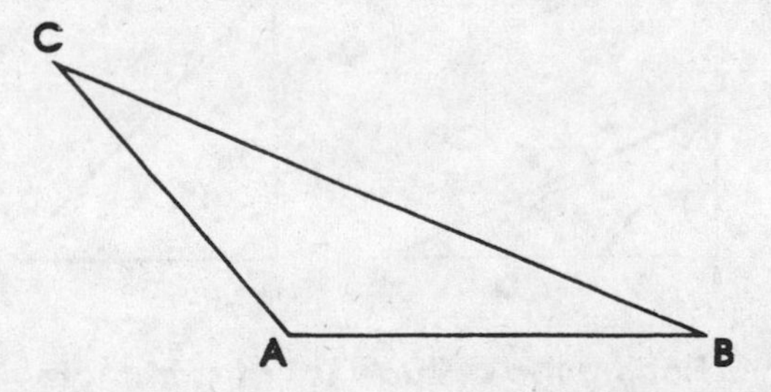

If one of the angles is a right angle, the triangle is called a *right* triangle.

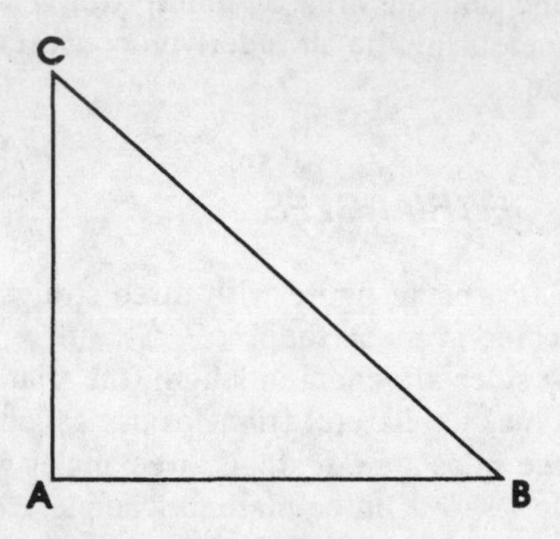

Each of the three corners of a triangle is called a *vertex* of the triangle. The bottom side of the triangle is called its *base,* and a line drawn perpendicular to the base from the top vertex is called its *altitude*. In the triangles below, the upper-case letters A, B, and C are at the vertexes. The line AB is the base, and the line CD is the altitude.

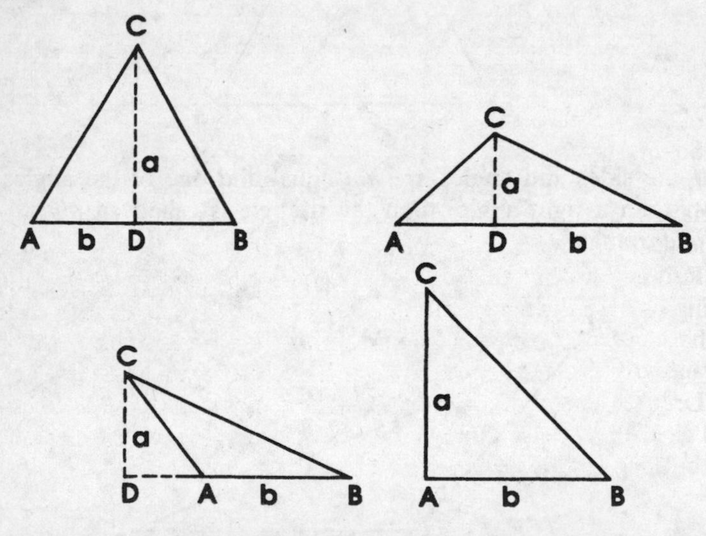

Finding the altitude (a) of a triangle

You can take any triangle, copy it exactly, and put the triangle and its copy together to form a parallelogram. This means that *any triangle forms half a parallelogram with the same base and altitude*!

Take a triangle . . .

and flip it over.

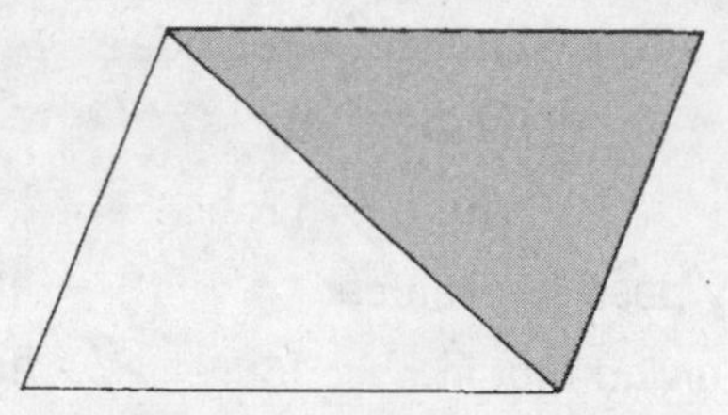

Now join the two to see a parallelogram!
The area of a parallelogram = altitude x base.
A triangle equals $\frac{1}{2}$ of a parallelogram.
Therefore, the area of a triangle = $\frac{1}{2}$ altitude x base.
$A = \frac{1}{2}\, a \times b$

So—how easy *is* it to find the area of a triangle? It's this easy:

If a triangle always forms half a parallelogram, then the area of a triangle is equal to half the area of a parallelogram having the same base and altitude as the triangle.

Remember that the area of a parallelogram is found by multiplying its base times its altitude. The area of a triangle, then, is half of that, or ½ of the base times the altitude. To find this, first multiply the base by the altitude, and then divide the product by 2.

Let's take any odd triangle—say, one shaped like the one below, and measure its base and its altitude. If our triangle is 14 feet across the bottom and its altitude is 12 feet, what is its area?

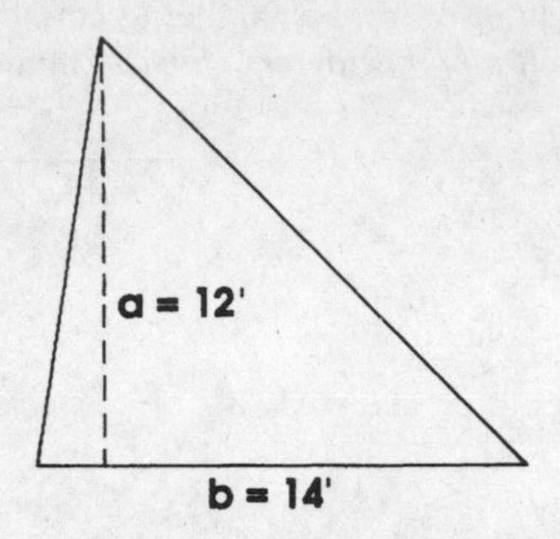

$$\text{Area} = \tfrac{1}{2}\ \text{base} \times \text{altitude}$$

$$\text{Base} = 14$$

$$\text{Altitude} = 12$$

$$\text{Multiply base} \times \text{altitude: } 14 \times 12 = 168$$

$$\text{Now divide that in half: } 168 \div 2 = 84$$

$$\text{Area} = 14 \times 12 = 168 \div 2 = \mathbf{84\ square\ feet}$$

Weird Fact About Triangles

Here's a piece of information to file in your memory for future reference: The sum of the three angles of any triangle always equals 180 degrees. If you are lucky enough to go on to geometry and trigonometry, you will use this fact. People in the construction trades make use of it all the time, as do surveyors, mapmakers, and many scientists.

The Amazing Properties of the Right Triangle

The queen of triangles is the right triangle. Its unusual characteristics make it a practical tool in many trades, particularly building, transportation, and surveying.

As you recall, the right triangle is defined by having one right angle in it. Since the sum of a triangle's angles is 180 degrees, that means the other two angles of a right triangle add up to 90 degrees.

The sides that make up the right angle are called the right triangle's *legs,* and the long side opposite the right angle is its *hypotenuse*.

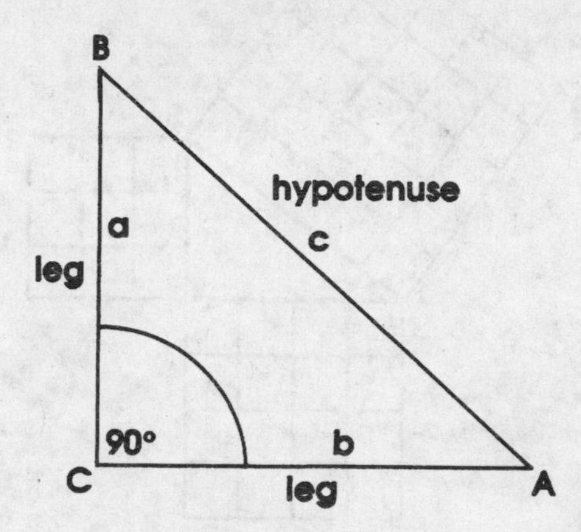

As you can see, if you set a right triangle on one of its legs, the other leg is always the same as its altitude. Therefore, you can always find the area of a right triangle by multiplying the legs and dividing the product in half.

$$\text{Area} = \tfrac{1}{2}(a \times b)$$

By the way, the standard way of drawing and lettering a right triangle designates the hypotenuse as side *c*—so that side *c* is opposite the right angle, C.

Here's a strange and useful fact about the right triangle: *The square of the hypotenuse equals the sum of the squares of the legs.*

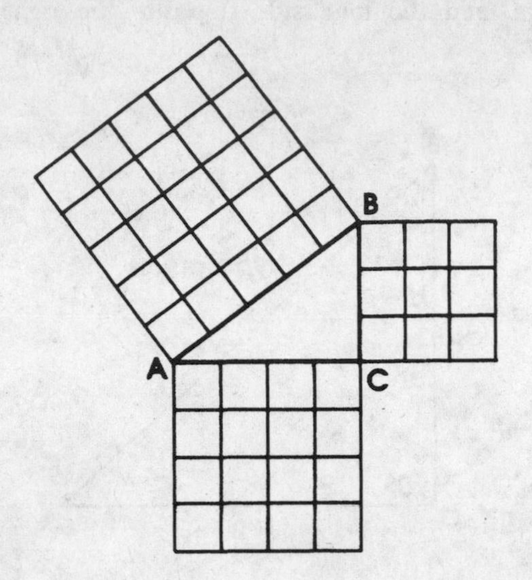

In this right triangle shown, leg *a* is 3 centimeters long, leg *b* is 4 centimeters long, and leg *c* is 5 centimeters long.

The square of *c* equals the square of *a* plus the square of *b*.

$$c^2 = a^2 + b^2$$

Does $5^2 = 3^2 + 4^2$?

$$3^2 = 9$$

$$4^2 = 16$$

$$9 + 16 = 25$$

$$5^2 = 25$$

$c^2 =$ **25 square centimeters**

It works. Imagine those squares as paper cutouts. Flip the 25-cm^2 box over so it still extends along the hypotenuse AB, but instead of extending above the triangle it folds back over it. Now draw a big square, which we'll mark DFHI, around this figure. Extend the leg AC out to point G and the leg BC down to point E.

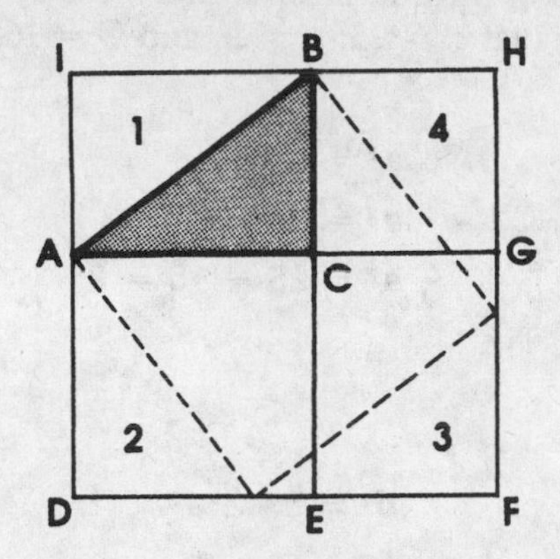

Now the large square DFHI is made up of the 4-cm² square and the 3-cm² square on the legs—they appear as square BCGH and square ACED—plus two equal rectangles, CEFG and ACBI. We see that one of these two equal rectangles, ACBI, divides evenly into two identical triangles, ABC and ABI. The four triangles formed around the outside of the big box—I've numbered them 1, 2, 3, and 4—are identical and they are equivalent to the two rectangles (if two of them equal one rectangle, four equal two rectangles).

Now, if you remove the four triangles from the big square, you're left with the dotted line square from the hypotenuse.

If you remove the two rectangles from the large square, you get the two squares on the legs AC and BC.

But we recall that the four triangles are the same size as the two rectangles. Therefore, the two remainders are also the same: The square along the hypotenuse equals the sum of the squares on the legs.

From this, it follows that the square of either leg of a right triangle equals the difference between the squares of the hypotenuse and the other leg.

$$c^2 = a^2 + b^2$$

$$a^2 = c^2 - b^2$$

$$b^2 = c^2 - a^2$$

Using our values of 5 for c, 3 for a, and 4 for b, let's see if that works. We know that $c^2 = 25$; $a^2 = 9$, and $b^2 = 16$.

$$a^2 = c^2 - b^2$$
$$a^2 = 5^2 - 4^2$$
$$a^2 = 25 - 16 = 9$$

That checks.

$$b^2 = c^2 - a^2$$
$$b^2 = 5^2 - 3^2$$
$$b^2 = 25 - 9 = \mathbf{16}$$

This strange relationship, first defined by a Greek mathematician and philosopher named Pythagoras about 500 B.C., plays an important part in the branch of math called *trigonometry*. It's called the *theorem of Pythagoras*, and the explanation or *proof* you see above is thought to have been invented by the ancient mathematician himself.

Two practical applications are widely used: (1) If you know the lengths of the legs, you can find the hypotenuse by squaring the figures you have, adding the results, and finding the square root of the sum; (2) If you know one leg and the hypotenuse, you can find the other leg by squaring the hypotenuse and the known leg and subtracting the results. The square root of the remainder is the missing leg.

THE AREA OF A TRAPEZOID

Back from the Twilight Zone! The trapezoid is a much more ordinary figure. Let's take a close look at it.

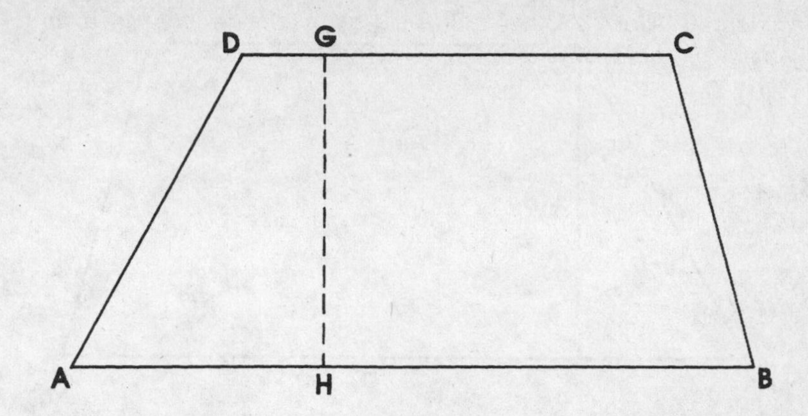

The two parallel sides are called the trapezoid's *bases*. The perpendicular distance between the two is its *altitude*. I've marked the altitude here with a dotted line between points G and H. The bases are AB and DC.

Now, if you draw a few extra dotted lines, you see that it's possible to regard the trapezoid as a shape formed by two triangles.

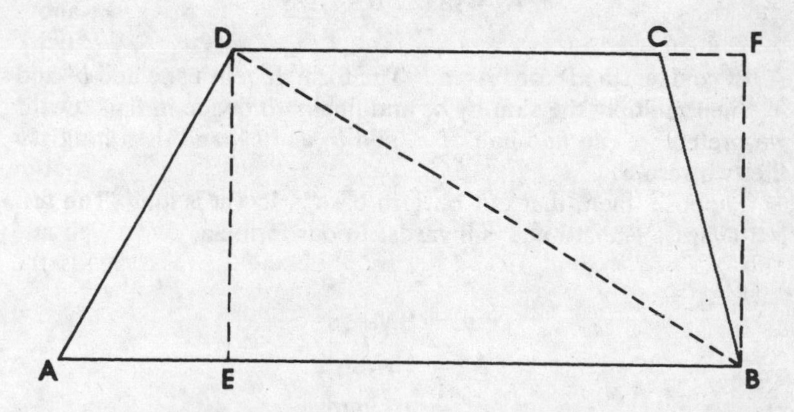

The triangles here are formed between the points ABD and BCD. And, as you can see, their altitudes—DE and BF—are the same as the trapezoid's. One triangle's base runs along the trapezoid's bottom base—AB—and the other runs along the trapezoid's top base—CD. So, to find the area of the trapezoid, we simply find the areas of the two triangles. To do that, find half the product of the trapezoid's altitude times the sum of its bases.

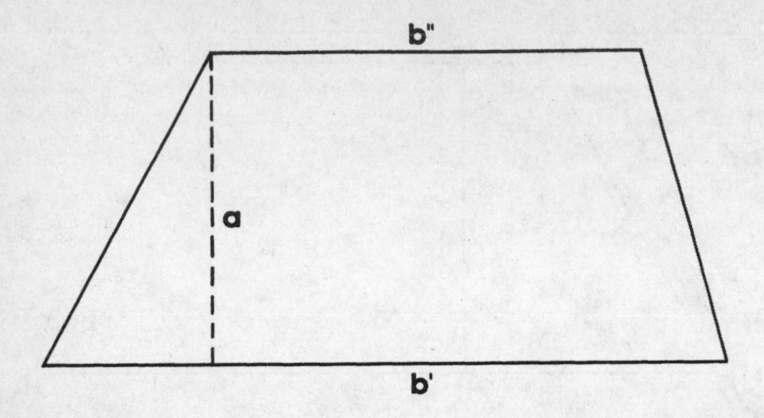

Let's call the trapezoid's altitude a. The bottom base is b' (referred to as "b prime") and the top base is b'' (pronounced "b second"). (See Figure 21.20.) To express our formula in symbols, we would write:

$$A = \tfrac{1}{2}a \times (b' + b'')$$

A, of course, stands for "Area." The formula tells us to add b' and b'', then multiply the sum by a, and finally to divide in half. Or, if we prefer, we can find half of a, add b' and b'', and then multiply the two results.

Suppose, then, that our bottom base is 15 yards long. The top is 10 yards. The altitude is 8 yards. In our formula,

$$a = 8 \text{ yards}$$

$$b' = 15 \text{ yards}$$

$$b'' = 10 \text{ yards}$$

$$A = \tfrac{1}{2}a \times (b' + b'')$$

$$A = \tfrac{1}{2} \text{ of } 8 \times (15 + 10) = 4 \times 25 = 100$$

$$A = \mathbf{100\ square\ yards}$$

THE PROPERTIES OF A CIRCLE

Another figure much loved by mathematicians is the circle. Like the triangle, it also implies some relationships that have many practical uses.

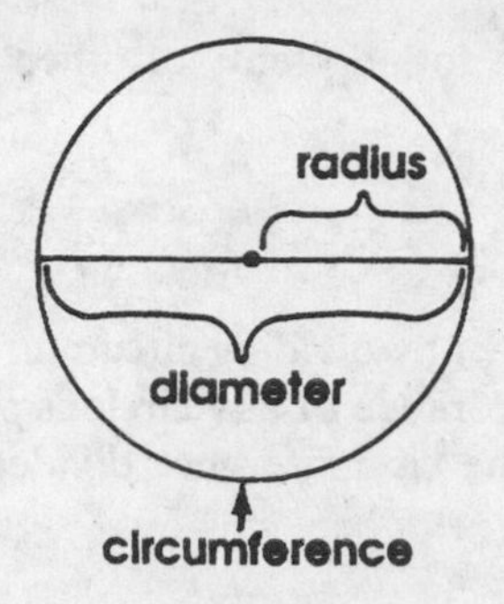

The outside line of a circle is called its *circumference.* Every point on the circumference is exactly the same distance from the *center.* A line passing through the center of the circle and touching the circumference on both ends is called the circle's *diameter.* A line from the center to the circumference is called its *radius* (plural, "radii"). And this leads us to the following well-known facts.

Weird Facts About Circles

If you measure the circumference and the diameter of *any* circle and then divide the circumference by the diameter, you will always get the same answer. It doesn't matter whether the circle is as big as a spaghetti plate or as small as your pinky ring. The magic number is approximately 3.1416—or about $3\frac{1}{7}$. Mathematicians call it *pi,* written with the Greek symbol π and pronounced "pie."

One other odd fact: As the circle is the symbol of a never-ending story (because it is a line that goes around and around without ending), so *pi* is a never-ending number!

The quotient is not an exact decimal, as far as anyone has been able to tell—and some fixated souls have divided it out to hundreds of places. To ten places, *pi* is 3.1415826535. It is always the same for every circle, no matter how far you divide the diameter into the circumference.

Finding an exact value for *pi* became a kind of obsession for number lovers in many cultures, because if you knew the precise value of *pi*, you would be able to find the dimensions of a square with the same area as that of a circle of any size. This goal was called "squaring the circle." In the year 1768, however, it was proven that *pi* has no exact value, and therefore it is impossible to square the circle.

However, we have found plenty of other practical applications for *pi*.

Using *pi* on the Circle

Because we found *pi* by dividing circumference by diameter, we know that the circumference of any circle is *pi* times the diameter, and the diameter is the circumference divided by *pi*. Or, to put it in math code:

$$C = \pi \times D$$

$$D = C \div \pi$$

C stands for "circumference" and D for "diameter."

If a circle has a diameter of 10 feet, the circumference is 3.1416 × 10, or 31.416 feet.

Since the diameter is twice the radius, we can find the circumference using either the diameter or the radius (*r*):

$$C = \pi \times D$$

$$C = 2 \times \pi \times r$$

In mathematical statements, we often leave the "times" symbol out to signify multiplication:

$$C = \pi D$$

$$C = 2\pi r$$

Two times *pi* is 6.2832. So if you know the radius, you can find the circumference easily by multiplying the radius times 6.2832. Similarly, if you know the circumference, the radius is the circumference divided by 6.2832.

$$r = C \div 2\pi$$

Now, to find the area of a circle, conjure up our enormous pizza pie. If we slice it into a lot of small pieces, we divide it into what are essentially triangles. The altitude of each pizza slice is the same as the pizza's radius. The sum of all their bases is the pizza's circumference. The sum of their areas is the area of the pizza.

Since the area of each slice is half the product of its base times its altitude, the sum of all the areas is the sum of all the products. The altitude is the same for all the slices, and the sum of the bases is the same as the pizza's circumference. So, the area of the pizza is one half the product of its radius and circumference.

$$A = \frac{1}{2} \times r \times C$$
$$A = \frac{1}{2} rC$$

Because the circumference is the same as 2C times the radius, we can also say that the area is 3.1416 times the radius times the radius, or 3.1416 times the radius squared.

$$A = \frac{1}{2}(2\pi \times r \times r)$$

$\frac{1}{2}$ of 2 = 1, so they cancel each other out. The shortened formula is:

$$A = \pi r^2$$

One of our pizza slices is four feet long (this is, after all, a *gigantic* pizza). So, we can apply this formula to find the area of the whole huge pizza pie:

$$A = \pi r^2$$

$A = 3.1416 \times 4^2 = 3.1416 \times 4 \times 4 =$ **50.2656 square feet**

If we had sliced the pizza in half and measured the length of that cut, we would have the diameter. Since the radius is half the diameter, the square of the radius is ¼ of the diameter squared. This means we can figure the area from the diameter by multiplying ¼ of C times the square of the diameter. A fourth of C is .7854, by the way. The diameter of this crazed pizza is 8 feet.

$$A = \pi/4 \times D^2$$

$$A = .7854 \times 8 \times 8 = \mathbf{50.2656\ square\ feet}$$

It's easy to reverse this process to find the diameter or the radius from a known area. To find the diameter, divide the area by .7854, and then find the square root of the quotient. The radius, then, is half the diameter. Or to find the radius directly, divide the area by 3.1416 and find the square root of the quotient.

$$D = A \div .7854$$

$$r = A \div 3.1416$$

Let's round off the area of our pizza to 50 square feet—some hungry soul, having grown impatient with our mathematical chat, has taken a couple of bites out of it.

$$D = A \div .7854$$

$$D = 50 \div .7854 = 63.66$$

$$D = 7.9 \text{ feet}$$

To find the radius, divide that in half: 3.85 feet. Or use the direct approach:

$$r^2 = A \div 3.1416$$

$$r^2 = 50 \div 3.1416 = 15.915$$

$$r = 3.85 \text{ feet}$$

The grass in the backyard has grown another inch since we began this conversation. And so, at last, now for the answer to the puzzle that started it all.

HOW BIG IS *THE "BACK FORTY"?*

This yard is oddly shaped. It started out as a rectangle, 35 feet deep by 40 feet long. Then somebody widened the driveway on the west side, cutting out a triangular slab. As a result, the yard is a trapezoid, 40 feet long on the south side, 32 feet on the north, and 35 feet on the east. The two corners on the west side are right angles.

Your kid sister felt it would be decorative to place two triangular flower beds in the northeast and northwest corners. She thought that they would be the same size if each ran 3 feet along the back fence and 4 feet along the side fence. She was wrong, but she was only five years old at the time. Your brother planted his favorite tree—hemlock—on the far side of the swimming pool. It occupies a tree well 5 feet across. The swimming pool itself is a rectangle, 20 feet long and 10 feet wide. A 3-foot-wide brick walkway extends 6 feet from the back door to the pool.

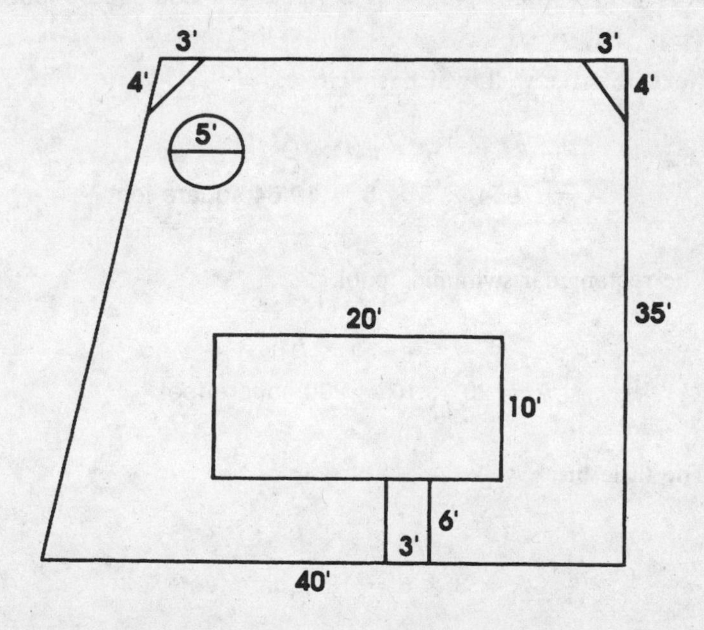

To figure how many square feet of grass you have to mow, we need first to find the area of the trapezoidal yard. Then we subtract the areas of all the landscaping elements that interrupt the lawn.

We have:

One trapezoid, 32 feet on the top, 40 feet on the bottom, and 35 feet perpendicularly from top to bottom
One circle, 5 feet in diameter
Two rectangles, one 20 feet × 10 feet and one 3 feet × 6 feet
One right triangle, whose legs are 3 feet and 4 feet long
One obtuse triangle, with a base of 4 feet and a leg of 3 feet

All *right!* To find the area of a trapezoid, we add the bases and multiply by half the altitude. Luckily, the 35-foot depth of the yard is the same as the trapezoid's altitude, since the east fence runs at right angles to the north and south fences.

$$A = \tfrac{1}{2} a \times (b' + b'')$$
$$A = \tfrac{1}{2}\,35 \times (32 + 40) = 17.5 \times 72 = 1{,}260 \text{ square feet}$$

Next, the tree well, which is a circle:

$$A = \pi/4 \times D^2$$
$$A = .7854 \times 5 \times 5 = 19.64 \text{ square feet}$$

The rectangular swimming pool:

$$A = a \times b$$
$$A = 20 \times 10 = 200 \text{ square feet}$$

The little brick walk:

$$A = a \times b$$
$$A = 3 \times 6 = 18 \text{ square feet}$$

The flower bed in the northwest corner, which is shaped like a right triangle:

$$A = \tfrac{1}{2}\,(a \times b)$$
$$A = \tfrac{1}{2}\,(3 \times 4) = 6 \text{ square feet}$$

The other flower bed, which is an obtuse triangle:

$$A = \tfrac{1}{2}\,(\text{base} \times \text{altitude})$$

We have to find the altitude for this triangle.

First, some fieldwork. Get a compass or a protractor (a plastic template showing the degrees of a half circle) and measure the angle formed by the northwest corner of the yard. Let's say it is 100°. With that information in mind, we can draw a scale model of the flower bed.

Let's take the 4-inch side as the base. Use a metric ruler, and count 1 centimeter as 1 foot. Draw a 4-centimeter base. At the left end, draw a line that meets the base at a 100° angle. Now, for a feeling of completeness, draw the third side of the triangle.

Extend a dotted line out from the left of the base. From the top vertex, drop an altitude line at right angles to the base. Measure its length, from the vertex to the extended baseline. I get about 2.5 centimeters. And now we have something to work with. Since each centimeter equals one foot, the altitude of the obtuse triangle-shaped garden is 2.5 feet.

$$A = \tfrac{1}{2}\,(\text{base} \times \text{altitude})$$
$$A = \tfrac{1}{2}\,(2.5 \times 4) = 5 \text{ square feet}$$

The rest is simplicity itself. Add up the areas of the shapes inside the yard:

Pool	200	square feet
Walkway	18	
Tree well	19.64	
Northwest garden	6	
Northeast garden	5	
	248.64	square feet

Subtract the total from the trapezoid's area:

Yard	1,260.00	square feet
Ornaments	− 248.64	
	1,011.36	**square feet of grass run amok in July**

PART V

THE NEXT STEP

22 Introducing Algebra

EVEN THOUGH YOU HAVEN'T been formally introduced, you've already met algebra. In the last chapter, when we were figuring the areas of various figures, we used statements like $A = \pi r^2$ and $A = \frac{1}{2} a \times (b' + b'')$. When we were explaining how to stun your associates by revealing the day of the week they were born, we came up with an expression that said

$$\text{Day of the week} = \frac{\text{year} + (\text{year} \div 4) + \text{day} + \text{SV}}{7}$$

If you decided that these strategies, which are really algebraic formulas, were sort of easy and kind of fun, then algebra has already wormed its way into your affections. There's nothing mysterious or weird about algebra. It's just another way of describing the thoughts we've already begun to express. That's all math is: a way of expressing certain kinds of thoughts.

In algebra, numbers may be represented by letters, which are used to solve problems by creating and playing with equations. Algebra uses negative numbers in addition to positive numbers, something you have already learned about. The use of powers—such as 3^3—is associated with algebra, as are "complex numbers"—such as $3 + (4 \times \sqrt{-1})$.

The word "algebra" comes from the Arabic *al-jebr,* which means "the reuniting of broken parts or bones." By extension, it implies the equating of like to like. As one writer remarks, it suggests the fitting together of broken bones inside limbs, where the surgeon could not see but had to work with unknowns. There is more than one kind of algebra. The garden-variety algebra you have already begun to learn is called "elementary algebra." Other types, such as Boolean algebra, may deal with special mathematical entities. In any case, all forms of algebra are characterized by the use of letters to represent *variables.*

A variable is a number whose value may change, as opposed to a *constant,* whose value is always the same. In the formula $A = \pi r^2$, we know that π is a constant. Its value is always 3.1416. The area of a circle (A) could be anything, and so could its radius (r); thus A and r are variables.

In algebra, some procedures are written differently from the way we write them in arithmetic. Let's survey the symbolism of algebra:

Suppose x stands for any number, and y stands for any number, which may or may not be the same as x.

When we write	We mean:
$x + y$	the sum of x and y (read it "x plus y")
$x - y$	the result of subtracting y from x ("x minus y")
xy	x times y (read it "xy")
$x \cdot y$	x times y ("x dot y"); raised dot may be used to indicate multiplication, to avoid confusion
x/y	x divided by y ("x over y")
$\frac{x}{y}$	x divided by y ("x over y")
$x \div y$	x divided by y
x^n	x raised to the power represented by the letter n ("x to the nth power")
$x = y$	x equals y

$x > y$	x is greater than y
$x < y$	x is less than y
$x \geq y$	x is equal to or greater than y
$x \leq y$	x is equal to or less than y
$\pm y$	plus or minus y
$(x + y)$	a number that is the sum of x and y
(xy)	a number that is the product of x and y
$(x - y)$	a number that is the remainder of x minus y

Parentheses are used to group figures together; when you see them, they mean you should perform those operations first, and then proceed with the rest of the formula. Normally, all multiplications and divisions are done first, followed by additions and subtractions. So, if you saw something like this:

$$(x + 6)\,(a^2 + b^2 + c^2) - 36y$$

Then, if you know the values of the variables represented by the letters, you can fill them in. Say that

$$x = 2$$
$$y = 4$$
$$a = 3$$
$$b = 5$$
$$c = 1$$

$$(x + 6)\,(a^2 + b^2 + c^2) - \overset{6}{36}y$$
$$(2 + 6)\,(3^2 + 5^2 + 1^2) - 36 \times 4$$

Normally, you would perform the arithmetic calculations in a certain order: first multiply, then divide, then add, then subtract. But whenever letters and figures are grouped inside parentheses, you should do the calculations inside the parentheses first. So, let's start there:

$$2 + 6 = 8$$

$$3^2 = 9;\ 5^2 = 25;\ 1^2 = 1;\ 9 + 25 + 1 = 35$$

Now you have

$$8 \times 35 - 36 \times 4$$

Again, first multiply:

$$8 \times 35 = 280$$
$$36 \times 4 = 144$$

And finally, subtract:

$$280 - 144 = \mathbf{136}$$

The best thing about algebra is that your variables don't have to be nailed down to one value or another for your statement to make sense. Through logic and common sense, you can identify relationships between numbers that *always* hold true. The equation $A = \pi r^2$ is just one example. Another is Einstein's famous statement of the theory of relativity, $E = mC^2$: Energy equals mass times the speed of light squared. No matter what figure you put in the place of m, the potential energy represented by that mass equals the mass times the square of the speed of light.

Algebra, then, is a tool that helps you deduce and express certain truths. Learning algebra trains you to think logically. This is why advanced math courses form an important part of any educated person's background.

Now You Can Do It

Translate these statements into algebra:
One number is added to a second number.
Multiply an unknown variable by three.
Three hundred forty-eight less 27.
Twenty-two is larger than 11.
Twenty-two is 2 times another number.
Take the result of subtracting 47 from 93 and divide it by 12.
First add 3 plus y; then subtract 5 from y; and then multiply the two results.

THE GROUND RULES

Algebra has a few basic rules, which are similar to those of arithmetic. For example:

- The sum of a series of numbers is the same, no matter in what order they are added.

$$(x + y) + z = x + (y + z) = x + y + z$$
$$x + y + z = z + y + x$$

- The product of numbers is the same, no matter in what order they are added.

$$xy = yz$$
$$(xy)z = x(yz) = xyz = zyx = yxz$$

- If you multiply a sum by a common factor, you get the same result that you would by multiplying each term contained in the sum by the same common factor.

$$(x + y)z = xz + yz$$

- You can subtract any number from any other number. This means, as we have seen, that the remainder is sometimes a negative number.
- You can divide any number by any other number, except 0.
- The number 0 is unique. Any number subtracted from itself equals zero, and any number multiplied by 0 equals 0. To put that algebraically, for every number, *a*:

$$a - a = 0$$
$$a \cdot 0 = 0$$

- The number 1 is unique. Any number (other than 0) divided by itself equals 1. That is:

$$a \div a = 1$$

And any number divided by 1 equals itself:

$$a \div 1 = a$$

- Any positive number multiplied by a negative number becomes a negative number; any negative number multiplied by another negative number becomes a positive number. For example:

$$a \cdot -1 = -a$$
$$-a \cdot -1 = a$$
$$1 \cdot -1 = -1$$
$$-1 \cdot -1 = 1$$

- Equal quantities can be added to equals, subtracted from equals, multiplied by equals, and divided by equals (except that division by 0 is not permitted), and the results will be equal. That is, if x equals y and z is another number, then:

$$x + z = y + z$$
$$x - z = y - z$$
$$xz = yz$$
$$x/z = y/z$$

Now You Can Do It

You have four numbers—*a, b, c,* and 14. Show how many ways they can be multiplied.

If $a = 5$ and $b = 15$, what is the result of $ab \div 75$?

Show another way of finding the answer to $x(14 + c)$.

If $x = 6$ and $y = 12$, what is the result of $xy \div 3x - y$?

If $x = 6$ and $y = 12$, what will happen if you multiply $(x + y)$ by -1?

MORE ON POWERS

With algebra, you can describe a number of operations that can be done with numbers raised to any power.

We've already talked about squaring and cubing numbers, and even touched on raising numbers to higher powers. Using algebraic symbols, you attach the exponent to a letter, just as you would for a number:

x^2 is read "*x* squared"

x^3 "*x* cubed"

x^5 "*x* to the fifth power"

x^n "*x* to the nth power"; n is used as a variable, like *x*

It's possible to have a power of zero, and a power of a fraction.

x^0 equals 1 for all values of *x* except 0

x^1 always equals *x*

x^{-n} *x* raised to any negative power is the same as 1 divided by a^n; in other words, $x^{-n} = 1/a^n$

x^{-1} always equals $1/x$. It's called the *reciprocal* of the number

$x^{1/2}$ always equals the square root of x; that is, $x^{1/2} = \sqrt{x}$

$x^{1/n}$ always equals the nth root of x; $x^{1/n} = \sqrt[n]{x}$

To multiply powers of the same number together, you add the exponents. For example, a^2 times $a^3 = a^5$. This is pretty obvious: $a^2 = a \cdot a$, and $a^3 = a \cdot a \cdot a$. Run them all together and you get $a \cdot a \cdot a \cdot a \cdot a$, or a^5.

Similarly, to divide powers of the same number, subtract the exponents. Thus, $a^4 \div a^2 = a^2$.

ALGEBRAIC EXPRESSIONS AND EQUATIONS

An *expression* is just a way of saying something in mathematical symbols. For example, in ordinary English we would say something like "10 less than b " The corresponding algebraic expression would say

$$b - 10$$

To say "20 more than b," we would write

$$b + 20$$

Expressions that have letters in them are called *literal* expressions.

When two expressions equal each other, then you have an *equation.* If we know that a is 10 less than b, for example, we can write

$$a = b - 10$$

And suppose we find out that the sum of b and 20 is *30*. We could express that as an equation, too:

$$b + 20 = 30$$

If we happened to know that a is ½ of b, then we would say that

$a = b/2$ "*a* equals *b* divided by 2"

$2a = b$ "2 times *a* equals *b*"

Given the figures in the previous equation, we could deduce that

$2a + 20 = 30$ "two times *a* plus 20 equals 30"

Each part of the equation is called a *term*. The terms of the equation above are $2a$, 20, and 30.

Now You Can Do It

What is the product of $2x^3$ times $14x^7$?
Divide $32c^6$ by $8c^2$.
Multiply $26a^2$ times $4(a^5 + y)$.

KEEPING EQUATIONS IN BALANCE

You can think of an equation as a scale that's in balance. You can do things to one side or another of it—add or subtract things, for example—but to keep it level, you have to do the same thing to both sides.

Suppose, then, that we have a scale such as the one the statue of Justice holds in her hand. We set a five-pound weight on the left side. The left side of the scale immediately drops. When we set another five pounds on the right side, the left side rises back to the level of the other side, and the scale is in balance.

Now, think of this equation as a balance pivoting on the equal sign:

$$5x = 2y$$

If we add 5 to the left term, we will throw the equation out of balance. And so, to keep it level, we must add 5 to *both* sides. Like this:

$$5x + 5 = 2y + 5$$

Let's experiment to see whether that seems to work. Let $x = 4$ and $y = 10$:

$$5x = 2y$$
$$5 \times 4 = 2 \times 10$$
$$20 = 20$$

That seems to be true. If we add 5 to one side, we get

$$5x + 5 = 2y$$
$$5 \times 4 + 5 = 2 \times 10?$$
$$20 + 5 = 20???$$

Nope! Add 5 to both sides:

$$5x + 5 = 2y + 5$$
$$5 \times 4 + 5 = 2 \times 10 + 5$$
$$20 + 5 = 20 + 5$$
$$25 = 25$$

This works for all the arithmetic operations. You can add to, subtract from, multiply, or divide *both sides* of an equation without throwing it out of whack.

Given what seems like an obvious piece of information, we can use it to manipulate equations and to *solve* them.

Now You Can Do It

Three pens cost $1.15. We can express this as an algebraic formula by letting p stand for *pens:*

$$3p = \$1.15$$

Now write a formula to show how much 1 pen costs (hint: divide both sides by 3). And how could you calculate how much each pen costs if 15 pens cost $1.15?

ALGEBRAIC ADDITION

Here's an equation:

$$a - b = 8$$

Now, suppose we know that b is 4, but we don't know the value of a. And we do want to know. . . . First, let's substitute 4 for b.

$$a - 4 = 8$$

If we add 4 to both sides, two things will happen: (1) The equation will stay in balance; and (2) the 4 on the left side will be canceled out.

$$a - 4 + 4 = 8 + 4$$
$$a = 8 + 4$$
$$a = 12$$

We actually found a concrete answer to that one—a real number that we can picture! But in algebra you don't always end up with a numeral. Sometimes the best you can do is simplify an expression. For example, suppose you come up with something like this:

$$2b - 6a = 3x + 4a + 86y$$

If you add $6a$ to both sides, you can at least see the simplest way to find $2b$:

$$2b - 6a = 3x + 4a + 86y$$
$$2b - 6a + 6a = 3x + 4a + 6a + 86y$$
$$2b = 3x + 10a + 86y$$

As you observe, you can *combine* like members in an equation. We had two multiples of a in the right term, and we managed to turn them into one by adding them.

What if we subtracted $2b$ from both sides? Then we would be left with a negative term on the left side: $-6a$. The revamped equation would look like this:

$$-6a = 3x + 4a + 86y - 2b$$

When we take $2b$ away from the left side, we leave nothing from which to subtract $6a$. That is, we leave 0: Subtract a positive number from 0 and you get a negative number.

You can combine multiplied terms that are being added, too, if each contains a common factor. The common factor becomes the multiplier, and you add the other two factors separately. In ab and bc, for example, the common factor is b. So if you added them, you would get

$$ab + bc, \text{ or}$$
$$(a + c)b$$

ALGEBRAIC SUBTRACTION

Nowhere other than in algebra is it more obvious that subtraction is a reverse form of addition. Subtracting a positive number (we'll talk more about positive and negative numbers below) is the same as adding a negative number; and subtracting a negative number is the same as adding a positive number. Take, for example:

$$2b - 6a = 3x$$

Let's subtract a negative number from this, say, $-4a$:

$$2b - 6a - (-4a) = 3x - (-4a)$$

Two negatives, as it were, make a positive: $-(-4a)$ becomes $+4a$. You combine the terms on the left side, and change the subtraction sign on the right side to an addition sign:

$$2b - 2a = 3x + 4a$$

Consider this expression:

$$97x - (a - b + c^2)$$

The subtraction sign tells you to change the signs inside the parentheses.

$$97x - a + b - c^2$$

In other words:

$$97x - (a - b + c^2) = 97x - a + b - c^2$$

Subtraction works the same way as addition to help solve an equation.

$$a + 4 = x$$

We see that 4 is added to one term, and so we can move it over to the other side by subtracting. If we have the value of x, this maneuver would allow us to learn what a is:

$$a = x - 4$$

Does this work, in real-world numbers? Let's say that x is 6 and a is 2.

$$a + 4 = x \quad a = x - 4$$
$$2 + 4 = 6 \quad 2 = 6 - 4$$

ALGEBRAIC MULTIPLICATION

In algebra, when you multiply single terms you write them next to each other and simplify. For example:

$$ac \cdot bd = abcd$$
$$7xy^2z^5 \cdot 2yxz^2a^3 = 14x \cdot x \cdot y^2 \cdot y \cdot z^5 \cdot z^2 \cdot a^3$$
$$= 14x^2y^3z^7a^3$$

In combining the terms, we also have combined powers. Because y is the same as y^1, the product of y times y^2 is y^3. And z^5 times z^2 equals z^7.

To multiply expressions, you just multiply a term at a time:

$$(a + b)(c + d) = ac + ad + bc + bd$$
$$(a + b)(a - b) = a^2 + (-ab) + ab + (-b^2) = a^2 - b^2$$

The $-ab$ cancels out the ab, and adding a negative b^2 is the same as subtracting b^2.

$$(7x + 3y)(2x + qz) = 14x^2 + 7xqz + 6xy + 3yqz$$
$$= 14x^2 + x(7qz + 6y) + 3yqz$$

We simplified the terms $7xqz$ and $6xy$ by combining them in a new expression. We could do this because x appeared as a factor in both, so we took the x out, multiplied and added the other factors as indicated, and then multiplied the result by x.

ALGEBRAIC DIVISION

Like terms may cancel each other out in a division problem. To divide powers, as we have seen, you subtract;

$$\frac{14xy^2z^3}{3x^2z} = \frac{14}{3}x^{-1}y^2z^2$$

Thus, $x \div x^2 = x^{-1}$, and $z^3 \div z^2 = z$. With nothing to divide into it, y^2 remains unchanged.

It's also possible to divide algebraic expressions the long way. Treat the entire expression that functions as the divisor the same way you would a single arithmetical number:

$$\begin{array}{r|l} & a + 4 \\ \hline a + 3 & a^2 + 7a + 12 \\ & \underline{a^2 + 3a} \\ & \quad 4a + 12 \\ & \quad \underline{4a + 12} \\ & \qquad 0 \end{array}$$

A FEW MORE WORDS ABOUT SIGNED NUMBERS

You learned what positive and negative numbers mean in Chapter 8, where we discovered that you can subtract a larger number from a smaller number. We used the time-honored example for teaching this concept, the thermometer, which has a scale that goes below 0°.

Algebra deals more extensively with signed numbers. In fact, mathematicians have devised some rules for dealing with positive and negative numbers.

Whatever a number's sign, its numeral is called its *absolute value.* For example, +16 and −16 have the same absolute value, 16. There are 16 units in each figure.

When a number is written with no sign in front of it, it is assumed to be positive. That is, 6 is the same as +6.

To *add numbers that have the same sign,* add their absolute values and give the sum the common sign.

$$\begin{array}{r} 24 \\ +6 \\ \hline 30 \end{array} \qquad \begin{array}{r} -24 \\ -6 \\ \hline -30 \end{array} \qquad \begin{array}{r} 6x \\ 2x \\ \hline 8x \end{array} \qquad \begin{array}{r} -6x \\ -2x \\ \hline -8x \end{array}$$

To *add numbers that have different signs,* subtract the smaller absolute value from the larger and give the difference the sign of the *larger* absolute value.

$$-6 + 8 = 2$$

$$-8 + 6 = -2$$

$$-40x + 10x = -30x$$

$$-10x + 40x = 30x$$

To *subtract a negative number from a positive number,* or *to subtract a positive number from a negative number,* change the sign of the subtrahend and add.

$$15 - (-3) = 15 + 3 = 18$$

$$-15 + 3 = -12$$

$$2x - (-x) = 3x$$

$$-2x - (-x) = -1x \text{ or } -x$$

To *multiply two numbers with the same sign,* multiply the figures in the simplest way and give the product a positive sign.

$$-45 \times -2 = 90$$

$$-3x \cdot -4x = 12x^2$$

$$6x \cdot 5x = 30x^2$$

To *multiply two numbers with different signs,* give the product a negative sign.

$$-45 \times 2 = -90$$

$$-5y \cdot 7y = -35y^2$$

To *multiply three numbers,* first multiply two of the figures, and then multiply the product by the third number.

$$2 \times 3 \times -5 = 6 \times -5 = -30$$

$$5x \cdot 4x \cdot -3x = 20x^2 \cdot -3x = -60x^3$$

$$-2 \times -3 \times 5 = +6 \times 5 = 30$$

$$-2 \times -3 \times -5 = 6 \times -5 = -30$$

To *divide numbers with the same signs,* give the quotient a positive sign.

$$6 \div 3 = 2$$

$$-6 \div -3 = 2$$

$$14x^2 \div 7x = 2x$$

$$-14x^2 \div 7x = -2x$$

To *divide numbers with different signs,* give the quotient a negative sign.

$$6 \div -3 = -2$$
$$-6 \div 3 = -2$$
$$22x^2 \div -2x = -11x$$
$$-22x^2 \div 2x = -11x$$

Now You Can Do It

As I write this, a storm is rolling in. Temperatures have dropped 30 degrees in the past half-hour. If the high today was 80° but the thermometer had fallen 15 degrees before the clouds came in, what is the temperature now? What if the thermometer drops another 40 degrees between now and midnight?

Joe owes Oliver Boxankel $50. He has $20 in his checking account. How deep in the red is Joe?

Joe pays Oliver $20 toward the above debt. Now he has nothing left for groceries. "Oliver," he says, "Lend me another fifty dollars so I can buy enough to eat till payday." Oliver, a generous soul, agrees. Now how deep in the red has Joe gone? Express his debt as a negative number.

Oliver works out an outrageous deal with Joe. If Joe doesn't pay in full within a week, he will pay Oliver three times his current debt. The debt will multiply by three with each passing week. Joe is a hopeless deadbeat and of course he doesn't pay up. After three weeks, how much does Joe owe Oliver on the debt he accrued in the last paragraph?

Finally, after sixteen weeks, Joe gets the job of his dreams. With one paycheck he is able to pay Oliver the entire debt that has built up over the past sixteen weeks. Oliver wants to buy a television, on sale at $565.98 at his favorite discount warehouse. Will the amount Joe owes pay for Oliver's television?

INTO THE FUTURE WITH ALGEBRA

As you go on in this branch of math, you learn to solve blindingly complicated-looking equations. There is, for example, the solution to the quadratic equation $ax^2 + bx + c = 0$, which is

$$x = \frac{-b \pm \sqrt{b^2 - 4ac}}{2a}$$

If your friends think you're a math whiz after this book, wait till they see what you do next!

23 A Test of Your Human Calculating

Just for fun, let's put all your new math skills together and figure out these problems.

1. To 1,020, add 1,020. Then add 20, and 20 again. Now add 10, and 10 again. What's the total?

2. You are on your way to fabulous riches, but until you reach your destination, you need to economize on small luxuries such as . . . well, furniture. You need a lamp table to go next to the easy chair you inherited from Great-aunt Matilda. Down at the Interior Deco Gallery they're charging an out-of-the-question $486.99 for a circular table of the sort you want.

However, you have some cinder blocks in the backyard. For the cost of a length of fabric and a piece of plywood sawed in the shape of a circle, you can build a piece of furniture that will be Just the Thing.

You stack the cinder blocks next to the chair and balance the plywood tabletop on them. Now all that remains is to disguise this construct with a round, handmade tablecloth.

The tabletop is 2 feet in diameter and 2½ feet high. You will run a ⅛-inch hem around the bottom of the tablecloth. The dimestore is selling the fabric you crave at $2.49 a yard. It is 48 inches wide. If the sales tax is 4.5 percent, how much will it cost to complete this project?

3. Oliver Q. Boxankle started a business. He owned $\frac{2}{5}$ of the stock, and a silent partner, who bankrolled him, owned $\frac{3}{5}$. A year later, Boxankle sold $\frac{1}{4}$ of his stock. How much of the company does he still own?

4. How many minutes are there in the month of December?

5. A stock clerk divides a number by 6 and gets a result of 12. He made a mistake: He should have multiplied. What is the correct answer he was looking for?

6. Check your answer to the preceding question by casting out nines.

7. In 1819, Simón Bolívar founded the Republic of Colombia. At that time, it had an area of 996,385 square miles and included the modern-day countries of Colombia, Ecuador, Panama, and Venezuela. In 1829, Venezuela seceded, leaving Colombia with 644,235 square miles. A year later, when Ecuador seceded, Colombia had 468,405 square miles. Panama stayed with Colombia until 1903, when it seceded and left its parent country with today's area of 439,830 square miles.

How many years after the founding of the Republic of Colombia did each of the daughter countries secede?

What is the area of each daughter country?

8. "I go one step backward for every two steps forward," cries Mom in frustration. If you take her literally, how many steps will she have to make in order to walk the twenty paces from the front door to the mailbox?

9. Using your powers of estimation, decide which of these two stacks of figures comes to the larger sum.

987654321	123456789
87654321	12345678
7654321	1234567
654321	123456
54321	12345
4321	1234
321	123
21	12
1	1

10. As part of your community's campaign to conserve electricity, you're keeping an eye on your electric meter. On January 1, the meter read 7,385 kilowatt-hours (kwh). Over the next six months, your electric-meter readings looked like this:

January: 7,642 kilowatt-hours (kwh)
February: 7,890 kwh
March: 8,142 kwh
April: 8,385 kwh
May: 8,624 kwh
June: 8,865 kwh

What was the difference, in kilowatt-hours, between your months of highest and lowest monthly energy consumption?

How many kilowatt-hours did you use during the spring months? The first day of spring comes in March.

If the power company sells electricity at 27 cents per kwh, what was your highest power bill? What was your lowest bill? What was your average monthly bill? How much did you pay the power company for the months of January through June?

11. Jet Set Jane drove from New York City to Denver, a distance of 2,100 miles, in 30 hours. How fast was her average speed?

12. Thanksgiving is coming up. Turkey is selling for $.89 a pound. You plan to have 18 people to dinner. Allowing for ¾ pound per person, how large a turkey should you buy? How much will it cost? How much will each serving cost?

13. On the other hand, turkey is pretty stereotyped. What if instead you served steak, at $3.59 a pound? Because the steak you will serve has no bones, you can allow ½ pound per person. How much steak will you buy to serve 18, at what cost? What will be the cost per guest?

14. You decide on turkey, and, to be sure you have enough, you purchase an 18-pounder. It takes 20 minutes per pound to cook a turkey. How long (in hours and minutes) will your turkey need to roast?

15. You happen to be in England this Thanksgiving, which means you will have to roast your turkey in an oven calibrated to Celsius

rather than Fahrenheit temperatures. You know a turkey is supposed to cook at 350° F. To what heat do you set your English oven?

16. Luckily, the English butcher shop that sold you the turkey weighed it in pounds. However, to make the stuffing, you're following a recipe in a French cookbook. It uses metric weights. To figure out how much stuffing to use, you need to know how many kilograms your 18-pound turkey weighs.

17. This French stuffing recipe calls for 25 cl of white wine. Estimate about how much this is in cups.

18. All the conversion between Fahrenheit and Celsius and between English and metric is driving you to the eggnog bowl. Before you can have a belt of eggnog, however, you must mix rum into the eggnog mix at a ratio of 1:4. If you have 3 quarts of eggnog mix, how much rum will you need to render you incapable of converting another number from English to metric figures?

19. Workers at the Boxankle Widget Manufacturing Corporation get one working day of vacation for each month in the 12 months of employment preceding their vacation date. As incentive, management adds an extra day of vacation time for each 5 years a person stays on the job. This is Joe's 24th year with Boxankle Widget. If he starts his vacation on the Tuesday after this year's three–day Fourth of July weekend, how many days will he be away from the factory?

20. You need new flooring covering in your house. The carpet you have chosen retails for $26.99 a square yard, plus 4 percent tax. You need to carpet this space:

Bedroom: 14 feet long, 12 feet wide
Bedroom: 13 feet long, 11 feet wide
Living room: 20 feet long, 14 feet wide
Dining room: 12 feet long, 10 feet wide
Hall: 10 feet long, 4 feet wide

How much carpet do you need to purchase? How much will it cost you?

21. Olivia bought a dress on sale for 75 percent of the marked price. She paid $62.50 for it. How much was the original price?

22. Tom Jones's Television and Repair Shop did well this year. Business increased 35 percent. Last year, the store earned $235,000. How much did it make this year?

23. About how many liter bottles of soft drink can be filled from four 10-gallon cans of soft drink?

24. What percentage of a quart is a pint?

25. Write these as signed (+ or −) numbers:
30 degrees below zero, Celsius
a debt of $42
a deposit of $152.65 to your savings account
a gain of 14 pounds
dropping 5 floors in an elevator
20° north latitude
15° south latitude

26. How many whole numbers between 1 and 10 can you form by using the digit 4 four times—no less, no more—by adding, subtracting, multiplying, or dividing?

For example: $1 = 44/44$
$2 = 4/4 + 4/4$

If you include $\sqrt{4}$, you should be able to reach 18 whole numbers.
—Martin Gardner

27. Write an equation to express the solution to each of these conundrums:

A number increased by 5 gives 42.
The length of one side of a square whose perimeter is 16 inches.
A number decreased by 12 produces 18.
Christine is 5 years older than Peter. The sum of their ages is 45. Find their ages.
Separate 30 into two parts such that one part is four times the other.
The volume of a box equals its length times its width times its height.

28. Express algebraically:

The number of feet in y yards
The number of inches in f feet
The number of months in x years
The number of quarters in d dollars
The number of cups in p pints

29. Juan Garcia went to the grocery store and purchased the following items on sale:

3 cans green beans, 90 cents (regularly 37 cents a can)
10 pounds of potatoes, 59 cents (regularly 15 cents a pound)
2 pounds of T-bone steaks, $6.78 (regularly $3.90 a pound)
3 avocados, $1.20 (down from 50 cents apiece last week)
2 fresh lemons, 30 cents (regularly 20 cents apiece)

How much did Juan save off the regular prices by buying sale-priced items?

30. The bank charges Eldon 10 cents for every check he writes. What will be left after he writes three checks—one for $14.98, one for $52.67, and one for $30.00—on a balance of $268.45?

31. A pizza is divided into 10 slices. If Maria gets .4 of the pizza, how many pieces does she get? How do you express that as a fraction?

32. If n and m represent two numbers, write the following:

The sum of the two numbers
The difference between the two numbers
The product of the two numbers
The first number divided by the second
The sum of the two numbers divided by the difference of the two numbers times the second number
The first number squared times the second number cubed minus the product of both numbers

33. Combine similar terms:

$5a + 2b + a + 4b$
$1x + 2x + 3y + 4y$
$2n + 3m + n + p$
$3x - x + y + 3y$

34. The Elm Street Giants have 25 players on their baseball team. Four players are 22 years old. What percent of the team is this?

35. Pete and Emily fly out of San Francisco at 11:45 A.M. Pacific Standard Time. They arrive 4 hours later in New York City. What time do they land?

36. You drive your car to school 5 days a week. Round trip from school to your home is 24 miles. Your car gets about 18 miles to the gallon. Gas at your favorite filling station is 99 cents a gallon. How much gas does it take to carry you back and forth to school each week? How much did it cost you to drive to school this week? If an 8 percent tax is included in the price of gas, how much gas tax did you pay this week?

37. You hire a chauffeur-driven limousine to drive you back and forth to school. This limo company charges $.89 a mile, and the round trip is 37 miles. How much do you tip the driver each day?

38. Christopher Gotrocks hires the limo driver to take him from his mansion to his fifteen-room summer cabin. The road takes them 248 miles east. Then they make a sharp left and drive 310 miles north. By the time they get to Christopher's cabin, they have put 558 miles on the limo. How much driving would they have saved if they could have driven to the cabin in a straight line? Write an equation showing how to find the length of this line, and solve it.

39. Christopher pays the limo driver $496.62. Now it's time to figure the tip. Christopher didn't get rich by letting money slip through his fingers. In fact, he's a bit of a cheapskate. He rounds the driver's fee down to $495 instead of up to $500, and he pays a gratuity of 16⅔ percent. How much does Christopher tip the driver?

40. Your dad takes you and your teacher to lunch at a popular coffee shop. She orders a bacon, lettuce, and tomato sandwich ($4.25), a chocolate milkshake ($1.90), and a side of french fries ($1.25). You amaze her and make Dad pleased as punch by quickly calculating her tab and 20 percent tip. What are these figures?

41. Naturally, the conversation during this luncheon turns to higher powers. You continue to floor your teacher by casually revealing the answer to 9^8, which of course you do by combining lower powers. With what figure do you bring her to her knees?

42. For your parting shot, you remark on the cube root of 17,576. Your teacher is forever wrapped around your little finger and you instantly supplant your kid brother, the Brat, as the apple of your father's eye. What number accomplishes this marvel?

ANSWERS FOR CHAPTER 23

1. 2,100—not 3,000!
2. $36.43, if you buy two 7-foot strips of fabric.
3. 2/5 × 1/4 = 2/20 = 1/10, the total amount of the company's stock he sold.
   Boxankle still owns 3 times that amount: 1/10 × 3 = 3/10.
4. 44,640 minutes
5. 432
6. Check number of 31 = 4
   Check number of 24 = 6
   Check number of 31 × 24 = 6
     4 × 6 = 24; cast out nines to get 6
     6 = 6
   Check number of 744 = 6
   Check number of 60 = 6
   Check number of 44,640 = 18; cast out nines to get 0
   6 × 6 = 36; cast out nines to get 0
   0 = 0
7. Venezuela, 10 years; 352,150 square miles
   Ecuador, 11 years; 175,830 square miles
   Panama, 84 years; 28,575 square miles
8. 58
9. They add up to the same total: 1,083,676,269
10. a) 18 kwh
    b) 975 kwh
    c) $69.39; $64.53; $66.60; $339.60
    Monthly kwh consumption:

January:	257
February:	248
March:	252
April:	243
May:	239
June:	241

11. 70 miles per hour
12. 13½ pounds; $12.02; $0.67 a serving
13. 9 pounds; $32.31; $1.80 a serving
14. 6 hours
15. 176.6°
16. 8.18 kilograms

17. 1 cup. A liter is about a quart; 25 cl is a quarter of a liter; there are 4 cups in a quart, and so 1 cup is about a quarter of a liter.
18. ¾ quart
19. 25 days. Because Joe starts his vacation on the day after a holiday, he gets an extra 3 days away from the assembly line. Weekends do not count as working days. Joe gets 4 bonus days because of his 24 years with Boxankle Widget.
20. Divide the linear foot measures by 3 to get linear yards.

4.66 yd. × 4 yd.	=	18.64 square yards
4.33 yd. × 3.66 yd.	=	15.85 sq. yd.
6.66 yd. × 4.66 yd.	=	31.00 sq. yd.
4 × 3.33 yd.	=	13.32 sq. yd.
3.33 yd. × 1.33 yd.	=	4.43 sq. yd.
		83.24 sq. yd.

83 sq. yd. × $26.99 = $2,240.17
4% of $2,240.17 = 89.61
$2,329.78

21. $62.50 ÷ .75 = $83.33
22. $235,000 × 1.35 = $317,250
23. Approximately 160 liters
24. 50 percent
25. −30° C
−$42
+$152.65
+14 pounds
−5 floors
+20°
−15°
26. $3 = \frac{4+4+4}{4}$

$4 = 4(4-4) + 4$

$5 = \frac{(4 \times 4) + 4}{4}$

$6 = 4 + \frac{4+4}{4}$

$7 = \frac{44}{4} - 4$

$8 = 4 + 4 + 4 - 4$

$9 = 4 + 4 + \frac{4}{4}$

$10 = \frac{44 - 4}{4}$

$11 = \frac{44}{\sqrt{4} + \sqrt{4}}$

$12 = \frac{44 + 4}{4}$

$13 = \frac{44}{4} = \sqrt{4}$

$14 = 4 + 4 + 4 + \sqrt{4}$

$15 = \frac{44}{4} + 4$

$16 = 4 + 4 + 4 + 4$

$17 = (4 \times 4) + \frac{4}{4}$

$18 = (4 \times 4) + 4 - \sqrt{4}$

27. $x + 5 = 42$
    $x = 16/4$
    $x - 12 = 18$
    $(P + 5) + P = 45; P = 20; C = 25$
    $30 = x + 4x$
    $v = 1wh$

28. $3y$
    $12f$
    $12x$
    $4d$
    $2p$

29. Total at regular prices: $ 1.11
    1.50
    7.80
    1.50
    .40
    $12.31

    Total at sale prices: .90
    .59
    6.78
    1.20
    .30
    $9.77

Difference: $12.31 − $9.77 = $2.54

30. $14.98 + $52.67 + $30.00 = 97.65
    $97.65 + $.30 = $97.95
    $268.45 − $97.95 = $170.50

31. 4 slices; 2/5 of the pizza

32. $n + m$
    $n - m$

nm
n/m
$\frac{(n+m)}{(n-m)m}$
$n^2m^3 - nm$

33. $6a + 6b$
$3x + 7y$
$3n + 3m + p$
$2x + 4y$
34. 16 percent
35. 6:45 P.M.
36. $24 \times 5 = 120$ miles
120 miles ÷ 18 mph = 6.66 gallons per week
6.66 gallons × 99 cents = $6.59, cost of gas this week
6.59 × .08 = 53 cents gas tax
37. 20% of the daily charge of $33 is $6.60.
38. The roads from Christopher's mansion to his summer home form the altitude and the base of a right triangle. A straight line between the two houses would form the hypotenuse of this triangle. We know that the square of a right triangle's hypotenuse equals the sum of the squares of the other two sides.
Let the hypotenuse be c, the altitude be a, and the base be b.
$c^2 = a^2 + b^2$
$c^2 = 96{,}100 + 61{,}504 = 157{,}604$
$c = 396.99$ miles, or about 397 miles
Christopher and his chauffeur could have saved 558 − 397, or 161 miles, if they could have driven straight to the summer cabin.
39. Divide by 6 to find 16⅔%: $82.50.
40. $7.40 plus tip of $1.50
41. 43,046,721. The eighth power is the square of the fourth power, or the square of the square of the square: 81 × 81 × 6,561.
42. 26

PART VI

APPENDICES

Glossary

Acute angle: An angle of less than 90° and more than 0°.
Acute triangle: A triangle whose angles are all acute.
Addend: One of two or more numbers to be added together.
Altitude: The height of a figure, measured perpendicularly to the base.
Algebraic expression: A statement that expresses arithmetical processes or relations with letter symbols.
Amount: A sum to be paid back after interest is computed.
Analog clock: A clock with a circular face on which hours, minutes, and seconds are marked.
Area: The amount of space, expressed in square units, inside a plane figure.
Base: The line on which a plane figure rests.
Celsius: A system of measuring temperature in which water freezes at 0° and boils at 100°.
Centigrade: Celsius.
Circle: A perfectly round plane figure.
Circumference: The distance around a circle.
Constant: A value that is always the same; for example, π is a constant (3.1416).
Common denominator: A number into which several fractions' denominators can be divided evenly.
Compass: A device for drawing circles.

Cube: A three-dimensional figure with six equal square faces. The product of a number multiplied by itself two times; for example, 9 is the cube of 3 (3 × 3 × 3 = 9).

Cube root: A number, which multiplied by itself twice, yields another number; for example, 2 is the cube root of 8 (2 × 2 × 2 = 8).

Decimal fraction: A part of a whole expressed in tenths, hundredths, thousandths, etc.

Decimal point: The dot at the left of a decimal fraction.

Degree: A unit used in measuring angles. A circle contains 360 degrees. A unit of temperature, Fahrenheit, Celsius, or Kelvin. Degrees are indicated with this symbol: °.

Denominator: In a fraction, the number that appears beneath the line.

Derive a root: To find a square root, cube root, fourth root, etc., of a number.

Diameter: A straight line drawn through the center of a circle from one side to the other.

Digit: One of the figures in a number; any of the ten numbers from 0 to 9.

Digital clock: A clock that shows the time in lighted or liquid-crystal digits.

Dividend: A number into which another number is to be divided.

Divisor: A number to be divided into another number.

Equation: A statement that shows two quantities are equal.

Exponent: A small raised figure appearing after a number, which indicates how many times the number is to be multiplied to reach the indicated power.

Extremes: The first and last numbers in a proportion.

Factor: One or more numbers multiplied together to give a product.

Fahrenheit: A system of measuring temperatures in which water freezes at 32° above 0° and boils at 212°.

Gratuity: Tip.

Height: The distance from the top to the base of a figure.

Heptagon: A plane figure with seven sides and seven angles.

Hexagon: A plane figure with six sides and six angles.

Hypotenuse: In a right triangle, the side opposite the right angle.

Interest: Money paid for the privilege of borrowing or using money, usually expressed as a percentage of the amount borrowed or used.

International Date Line: A line passing between the North and South poles at 180° of longitude.

Kelvin: A system of measuring temperatures in which one unit equals one Celsius degree and absolute 0° equals −273° Celsius.

Latitude: The distance north or south of the earth's Equator, measured in degrees, minutes, and seconds.

Longitude: The distancc east or west of the Prime Meridian, measured in degrees, minutes, and seconds.

Lowest terms: The farthest possible reduction of the numerator and denominator of a fraction; for example, 1/2 is 6/12 reduced to its lowest terms.

Means: The two inside numbers in a proportion.

Minuend: A number from which another number is to be subtracted.

Minute: 1/60 of an hour, or 1/60 of a degree.

Multiplicand: A number to be multiplied by another number.

Multiplier: A number used to multiply another number.

Negative number: A number whose value is less than 0; it is preceded by a minus sign (−).

Numerator: In a fraction, the number that appears on top of the line.

Obtuse angle: An angle of more than 90° and less than 180°.

Obtuse triangle: A triangle containing an obtuse angle.

Octagon: A plane figure with eight sides and eight angles.

Parallelogram: A four-sided figure whose opposite sides are parallel.

Pentagon: A plane figure with five sides and five angles.

Percent: A value expressed in hundredths; indicated by the percent sign (%).

Perimeter: The distance around the outside of a plane figure.

Perpendicular: At right angles.

***Pi*:** The ratio between the circumference of a circle and the diameter, symbolized π. Its approximate value is 3.1416 or 22/7.

Place value: The value of a digit according to where it is located in a number.

Plane figure: A flat shape.

Polygon: A closed plane figure.

Positive number: A number whose value is more than zero; may be preceded by a positive sign (+).

Power: The number of times a number appears as a factor in another number, as indicated by an exponent.

Prime Meridian: A line extending from the North to the South poles through Greenwich, England; it marks 0° of longitude.

Principal: An amount of money borrowed or used, upon which interest is calculated.

Product: The result of multiplying two or more numbers.

Proportion: A relationship in which two ratios are shown to be equal.

Protractor: A device for measuring angles.

Quadrilateral: Any four-sided figure.

Quotient: The result of dividing one number by another.

Radical sign: A symbol indicating a number is to be reduced to an indicated root: $\sqrt{\ }$

Radicand: The number placed under the radical sign.

Ratio: A comparison of two like quantities.

Rectangle: A plane figure with four sides and four right angles.

Remainder: The result of subtracting. Also, in division, the amount left over when a number does not divide evenly into another.

Right angle: Any 90° angle.

Right triangle: A triangle containing a right angle.

Root index: A raised number inside the crook of a radical symbol; used to indicate roots other than square roots.

Round number: An approximate number.

Second: 1/60th of a minute of time; one 1/60th of a minute of angular measure.

Sexagon: A plane figure with six sides and six angles.

Signed number: A number with a positive (+) or negative (−) sign in front of it.

Significant figure: In estimating, a number that it is important to have accurate.

Square: A rectangle whose sides are equal in length. The product of a number multiplied by itself once; *to square*: to multiply a number by itself once (for example, $2 \times 2 = 4$).

Square root: A number that, multiplied by itself, yields another number; for example, 2 is the square root of 4.

Straight angle: A 180° angle—which is a straight line.

Subtrahend: A number that is to be subtracted from another.

Sum: The result of adding.

Tax: An amount charged by a government to citizens who use government services; usually expressed as a percentage of income, sales, property value, etc.

Terms: The numbers in a fraction.

Time zone: A geographical area in which the same time is used.

Tip: An amount of money given to a person providing a service, in recognition of good work; usually expressed as a percentage of the cost of the service.

Trapezoid: A quadrilateral having two parallel sides.

Variable: A quantity that may assume one of various values.

Vertex: The point where the sides of an angle intersect.

Volume: The quantity that a solid figure can hold.

Tables

Tables of Weights and Measures

Avoirdupois Weight

$27^{11}/_{32}$ grains (gr)	1 dram (dr)
16 drams	1 ounce (oz), 437½ grains
16 ounces	1 pound (lb), 256 drams, 7,000 grains
100 pounds	1 hundredweight (cwt), 1,600 ounces
112 pounds	1 long hundredweight (1cwt)
20 hundredweight	1 ton (t), 2,000 pounds
20 long hundredweight	1 long ton (t), 2,240 pounds

Troy Weight

24 grains (gr)	1 pennyweight (dwt)
20 pennyweights	1 ounce (oz t) 480 grains
12 ounces	1 pound (lb t), 240 pennyweights, 5,760 grains

Apothecaries' Weight

20 grains (gr)	1 scruple
3 scruples	1 dram, 60 grains
8 drams	1 ounce, 24 scruples, 480 grains
12 ounces	1 pound, 96 drams, 288 scruples, 5,760 grains

Tables of Weights and Measures (cont.)

Linear Measure

12 inches (in)	1 foot (ft)
3 feet	1 yard (yd), 36 inches
5½ yards	1 rod (rd), 16½ feet
40 rods	1 furlong (fur), 220 yards, 660 feet
1 mile	5,280 feet
8 furlongs	1 statute mile (mi), 1,760 yards, 5,280 feet
3 miles	1 league (l), 5,280 yards

Square Measure

144 square inches (sq in)	1 square foot (sq ft)
9 square feet	1 square yard (sq yd), 1,296 square inches
30¼ square yards	1 square rod (sq rd), 272¼ square feet
160 square rods	1 acre (A), 4,840 square yards
640 acres	1 square mile (sq mi), 3,097,600 square yards
36 square miles	1 township

Cubic Measure

1,728 cubic inches (cu in)	1 cubic foot (cu ft)
27 cubic feet	1 cubic yard (cu yd)
144 cubic inches	1 board foot
128 cubic feet	1 cord

Liquid Measure

4 gills (gi)	1 pint (pt)
2 pints	1 quart (qt), 8 gills
4 quarts	1 gallon (gal), 8 pints, 32 gills
31½ gallons	1 barrel (bbl), 126 quarts
2 barrels	1 hogshead (hhd), 63 gallons, 252 quarts

Dry Measure

2 pints (pt)	1 quart (qt)
8 quarts	1 peck (pk), 16 pints
4 pecks	1 bushel (bu), 32 quarts, 64 pints
105 quarts	1 barrel (bbl), dry measure, 7,056 cubic inches

Tables of Weights and Measures (cont.)

Angular and Circular Measure

60 seconds (″)	1 minute (′)
60 minutes	1 degree (°)
90 degrees	1 quadrant (quad)
180 degrees	1 straight angle
4 quadrants	1 circle

Nautical Measure

6 feet	1 fathom (fath)
100 fathoms	1 cable's length (ordinary)
120 fathoms	1 cable's length (U.S. Navy)
10 cable lengths	1 nautical mile
1 nautical mile	1.1515 statute miles
60 nautical miles	1 degree (deg or °)
1 knot	1 nautical mile per hour

The Metric System

Linear Measure

10 millimeters	1 centimeter
10 centimeters	1 decimeter
10 decimeters	1 meter
10 meters	1 decameter
10 decameters	1 hectometer
10 hectometers	1 kilometer

Square Measure

100 sq millimeters	1 sq centimeter
100 sq centimeters	1 sq decimeter
100 sq decimeters	1 sq meter
100 sq meters	1 sq decameter
100 sq decameters	1 sq hectometer
100 sq hectometers	1 sq kilometer

Cubic Measure

1000 cu millimeters	1 cu centimeter
1000 cu centimeters	1 cu decimeter
1000 cu decimeters	1 cu meter

Liquid Measure

10 milliliters	1 centiliter
10 centiliters	1 deciliter
10 deciliters	1 liter
10 liters	1 decaliter
10 decaliters	1 hectoliter
10 hectoliters	1 kiloliter

Weights

10 milligrams	1 centigram
10 centigrams	1 decigram
10 decigrams	1 gram
10 grams	1 decagram
10 decagrams	1 hectogram
10 hectograms	1 kilogram
10 kilograms	1 quintal
10 quintals	1 ton

Conversion Table—Metric and English Systems

centimeter	0.3937 inch
meter	39.37 inches (exactly)
square centimeters	.1549997 square inches
square meter	1.195985 square yards
hectare	2.47104 acres
cubic meter	1.3079428 cubic yards
liter	.264178 gallon
liter	1.05671 liquid quarts
liter	.908102 dry quart
hectoliter	2.83782 bushels
gram	15.432356 grains
kilogram	2.204622341 pounds, avoirdupois
inch	2.540005 centimeters
yard	.9144018 meter
mile	1.6094 kilometers
square inch	6.451626 square centimeters
square yard	.8361307 square meter
acre	.404687 hectare
cubic yard	.7645594 cubic meter
gallon	3.785332 liters
liquid quart	.94633 liter
dry quart	1.101198 liters
bushel	35.23833 liters
grain	.064798918 gram
pound, avoirdupois	.45359237 kilogram

AMOUNT OF $1.00 AT COMPOUND INTEREST

To see how much will be owed after a given number of periods (years, months, weeks, days), look up the time span under *n*. Then run your finger over to the percentage of compound interest column. Multiply that number times the principal. The result is the total owed at the end of the loan period.

This table serves only as an example. More complete and complex tables have been published and are available at your library or bookstore.

n	½%	1%	2%	3%	4%	5%	6%
1	1.005 0000	1.010 0000	1.020 0000	1.030 0000	1.040 0000	1.050 0000	1.060 0000
2	1.010 0250	1.020 1000	1.040 4000	1.060 9000	1.081 6000	1.102 5000	1.123 6000
3	1.015 0751	1.030 3010	1.061 2080	1.092 7270	1.124 8640	1.157 6250	1.191 0160
4	1.020 1505	1.040 6040	1.082 4322	1.125 5088	1.169 8586	1.215 5062	1.262 4770
5	1.025 2513	1.051 0100	1.104 0808	1.159 2741	1.216 6529	1.276 2816	1.338 2256
6	1.030 3775	1.061 5202	1.126 1624	1.194 0523	1.265 3190	1.340 0956	1.418 5191
7	1.035 5294	1.072 1354	1.148 6857	1.229 8739	1.315 9318	1.407 1004	1.503 6303
8	1.040 7070	1.082 8567	1.171 6594	1.266 7701	1.368 5690	1.477 4554	1.593 8481
9	1.045 9106	1.093 6853	1.195 0926	1.304 7732	1.423 3118	1.551 3282	1.689 4790
10	1.051 1401	1.104 6221	1.218 9944	1.343 9164	1.480 2443	1.628 8946	1.790 8477
11	1.056 3958	1.115 6684	1.243 3743	1.384 2339	1.539 4541	1.710 3394	1.898 2986
12	1.061 6778	1.126 8250	1.268 2418	1.425 7609	1.601 0322	1.795 8563	2.012 1965
13	1.066 9862	1.138 0933	1.293 6066	1.468 5337	1.665 0735	1.885 6491	2.132.9283
14	1.072 3211	1.149 4742	1.319 4788	1.512 5897	1.731 6764	1.979 9316	2.260 9040
15	1.007 6827	1.160 9690	1.345 8683	1.557 9674	1.800 9435	2.078 9282	2.396 5582
16	1.083 0712	1.172 5786	1.372 7857	1.604 7064	1.872 9812	2.182 8746	2.540 3517
17	1.088 4865	1.184 3044	1.400 2414	1.652 8476	1.947 9005	2.292 0183	2.692 7728
18	1.093 9289	1.196 1475	1.428 2462	1.702 4331	2.025 8165	2.406 6192	2.854 3392
19	1.099 3986	1.208 1090	1.456 8112	1.753 5061	2.106 8492	2.526 9502	3.025 5995
20	1.104 8956	1.220 1900	1.485 9474	1.806 1112	2.191 1231	2.653 2977	3.207 1355
21	1.110 4201	1.232 3919	1.515 6663	1.860 2946	2.278 7681	2.785 9626	3.399 5636
22	1.115 9722	1.224 7159	1.545 9797	1.916 1034	2.369 9188	2.925 2607	3.603 5374
23	1.121 5520	1.257 1630	1.576 8993	1.973 5865	2.464 7155	3.071 5238	3.819 7497
24	1.127 1598	1.269 7346	1.608 4372	2.032 7941	2.563 3042	3.225 0999	4.048 9346
25	1.132 7956	1.282 4320	1.640 6060	2.093 7779	2.665 8363	3.386 3549	4.291 8707
26	1.138 4596	1.295 2563	1.673 4181	2.156 5913	2.772 4698	3.555 6727	4.549 3830
27	1.144 1519	1.308 2089	1.706 8865	2.221 2890	2.883 3686	3.773 4563	4.822 3459
28	1.149 8726	1.321 2910	1.741 0242	2.287 9277	2.998 7033	3.920 1291	5.111 6867
29	1.155 6220	1.334 5039	1.775 8447	2.356 5655	3.118 6514	4.116 1356	5.418 3879
30	1.161 4001	1.347 8489	1.811 3616	2.427 2625	3.243 3975	4.321 9424	5.743 4912
40	1.220 7942	1.488 8637	2.208 0397	3.262 0378	4.801 0206	7.039 9887	10.285 7179
50	1.283 2258	1.644 6318	2.691 5880	4.383 9060	7.106 6834	11.467 3998	18.420 1543
60	1.348 8502	1.816 6967	3.281 0308	5.891 6031	10.519 6274	18.679 1859	32.987 6908
70	1.417 8305	2.006 7634	3.999 5582	7.917 8219	15.571 6184	30.426 4255	59.075 9302
80	1.490 3386	2.216 7152	4.875 4392	10.640 8906	23.049 7991	49.561 4411	105.795 9935
90	1.566 5547	2.448 6327	5.943 1331	14.300 4671	34.119 3333	80.730 3650	189.464 5112
100	1.646 6685	2.704 8138	7.244 6461	19.218 6320	50.504 9482	131.501 2578	339.302 0835

Squares, Cubes, Square Roots, Cube Roots of Numbers

No.	Square	Cube	Sq. Rt.	Cu. Rt.	No.	Square	Cube	Sq. Rt.	Cu. Rt.
0	0	0	0.0000000	0.0000000	6	73 96	636 056	.2736185	.4140050
1	1	1	1.0000000	1.0000000	7	75 69	658 503	.3273791	.4310476
2	4	8	.4142136	.2599210	8	77 44	681 472	.3808315	.4479602
3	9	27	.7320508	.4422496	9	79 21	704 969	.4339811	.4647451
4	16	64	2.0000000	.5874011	90	81 00	729 000	9.4868330	4.4814047
5	25	125	2.2360680	1.7099759	1	82 81	753 571	.5393920	.4979414
6	36	216	.4494897	.8171206	2	84 64	778 688	.5916630	.5143574
7	49	343	.6457513	.9129312	3	86 49	804 357	.6436508	.5306549
8	64	512	.8284271	2.0000000	4	88 36	830 584	.6953597	.5468359
9	81	729	3.0000000	.0800838	95	90 25	857 375	9.7467943	4.5629026
10	1 00	1 000	3.1622777	2.1544347	6	92 16	884 736	.7979590	.5788570
11	1 21	1 331	.3166248	.2239801	7	94 09	912 673	.8488578	.5947009
12	1 44	1 728	.4641016	.2894285	8	96 04	941 192	.8994949	.6104363
13	1 69	2 197	.6055513	.3513347	9	98 01	970 299	.9498744	.6260650
14	1 96	2 744	.7416574	.4101423	100	1 00 00	1 000 000	10.0000000	4.6415888
15	2 25	3 375	3.8729833	2.4662121	1	1 02 01	1 030 301	.0498756	.6570095
16	2 56	4 096	4.0000000	.5198421	2	1 04 04	1 061 208	.0995049	.6723287
17	2 89	4 913	.1231056	.5712816	3	1 06 09	1 092 927	.1488916	.6875481
18	3 24	5 832	.2426407	.6207414	4	1 08 16	1 124 864	.1980390	.7026694
19	3 61	6 859	.3588989	.6684016	105	1 10 25	1 157 625	10.2469508	4.7176940
20	4 00	8 000	4.4721360	2.7144176	6	1 12 36	1 191 016	.2956301	.7326235
1	4 41	9 261	.5825757	.7589242	7	1 14 49	1 225 043	.3440804	.7474594
2	4 84	10 648	.6904158	.8020393	8	1 16 64	1 259 712	.3923048	.7622032
3	5 29	12 167	.7958315	.8438670	9	1 18 81	1 295 029	.4403065	.7768562
4	5 76	13 824	.8989795	.8844991	110	1 21 00	1 331 000	10.4880885	4.7914199
25	6 25	15 625	5.0000000	2.9240177	11	1 23 21	1 367 631	.5356538	.8058955
6	6 76	17 576	.0990195	.9624961	12	1 25 44	1 404 928	.5830052	.8202845
7	7 29	19 683	.1961524	3.0000000	13	1 27 69	1 442 897	.6301458	.8345881
8	7 84	21 952	.2915026	.0365890	14	1 29 96	1 481 544	.6770783	.8488076
9	8 41	24 389	.3851648	.0723168	115	1 32 25	1 520 875	10.7238053	4.8629441
30	9 00	27 000	5.4772256	3.1072325	16	1 34 56	1 560 896	.7703296	.8769990
1	9 61	29 791	.5677644	.1413807	17	1 36 89	1 601 613	.8166538	.8909732
2	10 24	32 768	.6568542	.1748021	18	1 39 24	1 643 032	.8627805	.9048681
3	10 89	35 937	.7445626	.2075343	19	1 41 61	1 685 159	.9087121	.9186847
4	11 56	39 304	.8309519	.2396118	120	1 44 00	1 728 000	10.9544512	4.9324241
35	12 25	42 875	5.9160798	3.2710663	1	1 46 41	1 771 561	11.0000000	.9460874
6	12 96	46 656	6.0000000	.3019272	2	1 48 84	1 815 848	.0453610	.9596757
7	13 69	50 653	.0827625	.3322219	3	1 51 29	1 860 867	.0905365	.9731898
8	14 44	54 872	.1644140	.3619754	4	1 53 76	1 906 624	.1355287	.9866310
9	15 21	59 319	.2449980	.3912114	125	1 56 25	1 953 125	11.1803399	5.0000000
40	16 00	64 000	6.3245553	3.4199519	6	1 58 76	2 000 376	.2249722	.0132979
1	16 81	68 921	.4031242	.4482172	7	1 61 29	2 048 383	.2694277	.0265257
2	17 64	74 088	.4807407	.4760266	8	1 63 84	2 097 152	.3137085	.0396842
3	18 49	79 507	.5574385	.5033981	9	1 66 41	2 146 689	.3578167	.0527743
4	19 36	85 184	.6332496	.5303483	130	1 69 00	2 197 000	11.4017543	5.0657970
45	20 25	91 125	6.7082039	3.5568933	1	1 71 61	2 248 091	.4455231	.0787531
6	21 16	97 336	.7823300	.5830479	2	1 74 24	2 299 968	.4891253	.0916434
7	22 09	103 823	.8556546	.6088261	3	1 76 89	2 352 637	.5325626	.1044687
8	23 04	110 592	.9282032	.6342412	4	1 79 56	2 406 104	.5758369	.1172299
9	24 01	117 649	7.0000000	.6593057	135	1 82 25	2 460 375	11.6189500	5.1299278
50	25 00	125 000	7.0710678	3.6840315	6	1 84 96	2 515 456	.6619038	.1425632
1	26 01	132 651	.1414284	.7084298	7	1 87 69	2 571 353	.7046999	.1551367
2	27 04	140 608	.2111026	.7325112	8	1 90 44	2 628 072	.7473401	.1676493
3	28 09	148 877	.2801099	.7562858	9	1 93 21	2 685 619	.7898261	.1801015
4	29 16	157 464	.3484692	.7797631	140	1 96 00	2 744 000	11.8321596	5.1924941
55	30 25	166 375	7.4161985	3.8029525	1	1 98 81	2 803 221	.8743422	.2048279
6	31 36	175 616	.4833148	.8258624	2	2 01 64	2 863 288	.9163753	.2171034
7	32 49	185 193	.5498344	.8485011	3	2 04 49	2 924 207	.9582607	.2293215
8	33 64	195 112	.6157731	.8708766	4	2 07 36	2 985 984	12.0000000	.2414828
9	34 81	205 379	.6811457	.8929964	145	2 10 25	3 048 625	12.0415946	5.2535879
60	36 00	216 000	7.7459667	3.9148676	6	2 13 16	3 112 136	.0830460	.2656374
1	37 21	226 981	.8102497	.9364972	7	2 16 09	3 176 523	.1243557	.2776321
2	38 44	238 328	.8740079	.9578916	8	2 19 04	3 241 792	.1655251	.2895725
3	39 69	250 047	.9372539	.9790572	9	2 22 01	3 307 949	.2065556	.3014592
4	40 96	262 144	8.0000000	4.0000000	150	2 25 00	3 375 000	12.2474487	5.3132928
65	42 25	274 625	8.0622577	4.0207258	1	2 28 01	3 442 951	.2882057	.3250740
6	43 56	287 496	.1240384	.0412400	2	2 31 04	3 511 808	.3288280	.3368033
7	44 89	300 763	.1853528	.0615481	3	2 34 09	3 581 577	.3693169	.3484812
8	46 24	314 432	.2462113	.0816551	4	2 37 16	3 652 264	.4096736	.3601084
9	47 61	328 509	.3066239	.1015659	155	2 40 25	3 723 875	12.4498996	5.3716854
70	49 00	343 000	8.3666003	4.1212853	6	2 43 36	3 796 416	.4899960	.3832126
1	50 41	357 911	.4261498	.1408177	7	2 46 49	3 869 893	.5299641	.3946907
2	51 84	373 248	.4852814	.1601676	8	2 49 64	3 944 312	.5698051	.4061202
3	53 29	389 017	.5440037	.1793392	9	2 52 81	4 019 679	.6095202	.4175015
4	54 76	405 224	.6023253	.1983365	160	2 56 00	4 096 000	12.6491106	5.4288352
75	56 25	421 875	8.6602540	4.2171633	1	2 59 21	4 173 281	.6885775	.4401218
6	57 76	438 976	.7177979	.2358236	2	2 62 44	4 251 528	.7279221	.4513618
7	59 29	456 533	.7749644	.2543209	3	2 65 69	4 330 747	.7671453	.4625556
8	60 84	474 552	.8317609	.2726587	4	2 68 96	4 410 944	.8062485	.4737037
9	62 41	493 039	.8881944	.2908404	165	2 72 25	4 492 125	12.8452326	5.4848066
80	64 00	512 000	8.9442719	4.3088694	6	2 75 56	4 574 296	.8840987	.4958647
1	65 61	531 441	9.0000000	.3267487	7	2 78 89	4 657 463	.9228480	.5068784
2	67 24	551 368	.0553851	.3444815	8	2 82 24	4 741 632	.9614814	.5178484
3	68 89	571 787	.1104336	.3620707	9	2 85 61	4 826 809	13.0000000	.5287748
4	70 56	592 704	.1651514	.3795191	170	2 89 00	4 913 000	13.0384048	5.5396583
85	72 25	614 125	9.2195445	4.3968297	1	2 92 41	5 000 211	.0766968	.5504991
6	73 96	636 056	.2736185	.4140050	2	2 95 84	5 088 448	.1148770	.5612978

No.	Square	Cube	Sq. Rt.	Cu. Rt.	No.	Square	Cube	Sq. Rt.	Cu. Rt.
2	2 95 84	5 088 448	.1148770	.5612978	260	6 76 00	17 576 000	16.1245155	6.3825043
3	2 99 29	5 177 717	.1529464	.5720546	1	6 81 21	17 779 581	.1554944	.3906765
4	3 02 76	5 268 024	.1909060	.5827702	2	6 86 44	17 984 728	.1864141	.3988279
175	3 06 25	5 359 375	13.2287566	5.5934447	3	6 91 69	18 191 447	.2172747	.4069585
6	3 09 76	5 451 776	.2664992	.6040787	4	6 96 96	18 399 744	.2480768	.4150687
7	3 13 29	5 545 233	.3041347	.6146724	265	7 02 25	18 609 625	16.2788206	6.4231583
8	3 16 84	5 639 752	.3416641	.6252263	6	7 07 56	18 821 096	.3095064	.4312276
9	3 20 41	5 735 339	.3790882	.6357408	7	7 12 89	19 034 163	.3401346	.4392767
180	3 24 00	5 832 000	13.4164079	5.6462162	8	7 18 24	19 248 832	.3707055	.4473057
1	3 27 61	5 929 741	.4536240	.6566528	9	7 23 61	19 465 109	.4012195	.4553148
2	3 31 24	6 028 568	.4907376	.6670511	270	7 29 00	19 683 000	16.4316767	6.4633041
3	3 34 89	6 128 487	.5277493	.6774114	1	7 34 41	19 902 511	.4620776	.4712736
4	3 38 56	6 229 504	.5646600	.6877340	2	7 39 84	20 123 648	.4924225	.4792236
185	3 42 25	6 331 625	13.6014705	5.6980192	3	7 45 29	20 346 417	.5227116	.4871541
6	3 45 96	6 434 856	.6381817	.7082675	4	7 50 76	20 570 824	.5529454	.4950653
7	3 49 69	6 539 203	.6747943	.7184791	275	7 56 25	20 796 875	16.5831240	6.5029572
8	3 53 44	6 644 672	.7113092	.7286543	6	7 61 76	21 024 576	.6132477	.5108300
9	3 57 21	6 751 269	.7477271	.7387936	7	7 67 29	21 253 933	.6433170	.5186839
190	3 61 00	6 859 000	13.7840488	5.7488971	8	7 72 84	21 484 952	.6733320	.5265189
1	3 64 81	6 967 871	.8202750	.7589652	9	7 78 41	21 717 639	.7032931	.5343351
2	3 68 64	7 077 888	.8564065	.7689982	280	7 84 00	21 952 000	16.7332005	6.5421326
3	3 72 49	7 189 057	.8924440	.7789966	1	7 89 61	22 188 041	.7630546	.5499116
4	3 76 36	7 301 384	.9283883	.7889604	2	7 95 24	22 425 768	.7928556	.5576722
195	3 80 25	7 414 875	13.9642400	5.7988900	3	8 00 89	22 665 187	.8226038	.5654144
6	3 84 16	7 529 536	14.0000000	.8087857	4	8 06 56	22 906 304	.8522995	.5731385
7	3 88 09	7 645 373	.0356688	8186479	285	8 12 25	23 149 125	16.8819430	6.5808443
8	3 92 04	7 762 392	.0712473	8284767	6	8 17 96	23 393 656	.9115345	.5885323
9	3 96 01	7 880 599	.1067360	8382725	7	8 23 69	23 639 903	.9410743	.5962023
200	4 00 00	8 000 000	14.1421356	5.8480355	8	8 29 44	23 887 872	.9705627	.6038545
1	4 04 01	8 120 601	.1774469	.8577660	9	8 35 21	24 137 569	17.0000000	.6114890
2	4 08 04	8 242 408	.2126704	.8674643	290	8 41 00	24 389 000	17.0293864	6.6191060
3	4 12 09	8 365 427	.2478068	.8771307	1	8 46 81	24 642 171	.0587221	.6267054
4	4 16 16	8 489 664	.2828569	.8867653	2	8 52 64	24 897 088	.0880075	.6342874
205	4 20 25	8 615 125	14.3178211	5.8963685	3	8 58 49	25 153 757	.1172428	.6418522
6	4 24 36	8 741 816	.3527001	.9059406	4	8 64 36	25 412 184	.1464282	.6493998
7	4 28 49	8 869 743	.3874946	.9154817	295	8 70 25	25 672 375	17.1755640	6.6569302
8	4 32 64	8 998 912	.4222051	.9249921	6	8 76 16	25 934 336	.2046505	.6644437
9	4 36 81	9 129 329	.4568323	.9344721	7	8 82 09	26 198 073	.2336879	.6719403
210	4 41 00	9 261 000	14.4913767	5.9439220	8	8 88 04	26 463 592	.2626765	.6794200
11	4 45 21	9 393 931	.5258390	.9533418	9	8 94 01	26 730 899	.2916165	.6868831
12	4 49 44	9 528 128	.5602198	.9627320	300	9 00 00	27 000 000	17.3205081	6.6943295
13	4 53 69	9 663 597	.5945195	.9720926	1	9 06 01	27 270 901	.3493516	.7017593
14	4 57 96	9 800 344	.6287388	.9814240	2	9 12 04	27 543 608	.3781472	.7091729
215	4 62 25	9 938 375	14.6628783	5.9907264	3	9 18 09	27 818 127	.4068952	.7165700
16	4 66 56	10 077 696	.6969385	6.0000000	4	9 24 16	28 094 464	.4355958	.7239508
17	4 70 89	10 218 313	.7309199	.0092450	305	9 30 25	28 372 625	17.4642492	6.7313155
18	4 75 24	10 360 232	.7648231	.0184617	6	9 36 36	28 652 616	.4928557	.7386641
19	4 79 61	10 503 459	.7986486	.0276502	7	9 42 49	28 934 443	.5214155	.7459967
220	4 84 00	10 648 000	14.8323970	6.0368107	8	9 48 64	29 218 112	.5499288	.7533134
1	4 88 41	10 793 861	.8660687	.0459436	9	9 54 81	29 503 629	.5783958	.7606143
2	4 92 84	10 941 048	.8996644	.0550489	310	9 61 00	29 791 000	17.6068169	6.7678995
3	4 97 29	11 089 567	.9331845	.0641270	11	9 67 21	30 080 231	.6351921	.7751690
4	5 01 76	11 239 424	.9666295	.0731779	12	9 73 44	30 371 328	.6635217	.7824229
225	5 06 25	11 390 625	15.0000000	6.0822020	13	9 79 69	30 664 297	.6918060	.7896613
6	5 10 76	11 543 176	.0332964	.0911994	14	9 85 96	30 959 144	.7200451	.7968844
7	5 15 29	11 697 083	.0665192	.1001702	315	9 92 25	31 255 875	17.7482393	6.8040921
8	5 19 84	11 852 352	.0996689	.1091147	16	9 98 56	31 554 496	.7763888	.8112847
9	5 24 41	12 008 989	.1327460	.1180332	17	10 04 89	31 855 013	.8044938	.8184620
230	5 29 00	12 167 000	15.1657509	6.1269257	18	10 11 24	32 157 432	.8325545	.8256242
1	5 33 61	12 326 391	.1986842	.1357924	19	10 17 61	32 461 759	.8605711	.8327714
2	5 38 24	12 487 168	.2315462	.1446337	320	10 24 00	32 768 000	17.8885438	6.8399037
3	5 42 89	12 649 337	.2643375	.1534495	1	10 30 41	33 076 161	.9164729	.8470213
4	5 47 56	12 812 904	.2970585	.1622401	2	10 36 84	33 386 248	.9443584	.8541240
235	5 52 25	12 977 875	15.3297097	6.1710058	3	10 43 29	33 698 267	.9722008	.8612120
6	5 56 96	13 144 256	.3622915	.1797466	4	10 49 76	34 012 224	18.0000000	.8682855
7	5 61 69	13 312 053	.3948043	.1884628	325	10 56 25	34 328 125	18.0277564	6.8753443
8	5 66 44	13 481 272	.4272486	.1971544	6	10 62 76	34 645 976	.0554701	.8823888
9	5 71 21	13 651 919	.4596248	.2058218	7	10 69 29	34 965 783	.0831413	.8894188
240	5 76 00	13 824 000	15.4919334	6.2144650	8	10 75 84	35 287 552	.1107703	.8964345
1	5 80 81	13 997 521	.5241747	.2230843	9	10 82 41	35 611 289	.1383571	.9034359
2	5 85 64	14 172 488	.5563492	.2316797	330	10 89 00	35 937 000	18.1659021	6.9104232
3	5 90 49	14 348 907	.5884573	.2402515	1	10 95 61	36 264 691	.1934054	.9173964
4	5 95 36	14 526 784	.6204994	.2487998	2	11 02 24	36 594 368	.2208672	9243556
245	6 00 25	14 706 125	15.6524758	6.2573248	3	11 08 89	36 926 037	.2482876	.9313008
6	6 05 16	14 886 936	.6843871	.2658266	4	11 15 56	37 259 704	.2756669	.9382321
7	6 10 09	15 069 223	.7162336	.2743054	335	11 22 25	37 595 375	18.3030052	6.9451496
8	6 15 04	15 252 992	.7480157	.2827613	6	11 28 96	37 933 056	.3303028	.9520533
9	6 20 01	15 438 249	.7797338	.2911946	7	11 35 69	38 272 753	.3575598	.9589434
250	6 25 00	15 625 000	15.8113883	6.2996053	8	11 42 44	38 614 472	.3847763	.9658198
1	6 30 01	15 813 251	8429795	.3079935	9	11 49 21	38 958 219	.4119526	.9726826
2	6 35 04	16 003 008	8745079	.3163596	340	11 56 00	39 304 000	18.4390889	6.9795321
3	6 40 09	16 194 277	.9059737	.3247035	1	11 62 81	39 651 821	.4661853	.9863681
4	6 45 16	16 387 064	.9373775	.3330256	2	11 69 64	40 001 688	.4932420	.9931906
255	6 50 25	16 581 375	15.9687194	6.3413257	3	11 76 49	40 353 607	.5202592	7.0000000
6	6 55 36	16 777 216	16.0000000	.3496042	4	11 83 36	40 707 584	.5472370	.0067961
7	6 60 49	16 974 593	.0312195	.3578611	345	11 90 25	41 063 625	18.5741756	7.0135791
8	6 65 64	17 173 512	.0623784	.3660968	6	11 97 16	41 421 736	.6010752	.0203490
9	6 70 81	17 373 979	.0934769	.3743111	7	12 04 09	41 781 923	.6279360	.0271058
260	6 76 00	17 576 000	16.1245155	6.3825043	8	12 11 04	42 144 192	.6547581	.0338497

No.	Square	Cube	Sq. Rt.	Cu. Rt.	No.	Square	Cube	Sq. Rt.	Cu. Rt.
8	12 11 04	42 144 192	.6547581	.0338497	6	19 00 96	82 881 856	.8806130	.5827865
9	12 18 01	42 508 549	.6815417	.0405806	7	19 09 69	83 453 453	.9045450	.5885793
350	12 25 00	42 875 000	18.7082869	7.0472987	8	19 18 44	84 027 672	.9284495	.5943633
1	12 32 01	43 243 551	.7349940	.0540041	9	19 27 21	84 604 519	.9523268	.6001385
2	12 39 04	43 614 208	.7616630	.0606967	440	19 36 00	85 184 000	20.9761770	7.6059049
3	12 46 09	43 986 977	.7882942	.0673766	1	19 44 81	85 766 121	21.0000000	.6116626
4	12 53 16	44 361 864	.8148877	.0740440	2	19 53 64	86 350 888	.0237960	.6174116
355	12 60 25	44 738 875	18.8414437	7.0806988	3	19 62 49	86 938 307	.0475652	.6231519
6	12 67 36	45 118 016	.8679623	.0873411	4	19 71 36	87 528 384	.0713075	.6288837
7	12 74 49	45 499 293	.8944436	.0939709	445	19 80 25	88 121 125	21.0950231	7.6346067
8	12 81 64	45 882 712	.9208879	.1005885	6	19 89 16	88 716 536	.1187121	.6403213
9	12 88 81	46 268 279	.9472953	.1071937	7	19 98 09	89 314 623	.1423745	.6460272
360	12 96 00	46 656 000	18.9736660	7.1137866	8	20 07 04	89 915 392	.1660105	.6517247
1	13 03 21	47 045 881	19.0000000	.1203674	9	20 16 01	90 518 849	.1896201	.6574138
2	13 10 44	47 437 928	.0262976	.1269360	450	20 25 00	91 125 000	21.2132034	7.6630943
3	13 17 69	47 832 147	.0525589	.1334925	1	20 34 01	91 733 851	.2367606	.6687665
4	13 24 96	48 228 544	.0787840	.1400370	2	20 43 04	92 345 408	.2602916	.6744303
365	13 32 25	48 627 125	19.1049732	7.1465695	3	20 52 09	92 959 677	.2837967	.6800857
6	13 39 56	49 027 896	.1311265	.1530901	4	20 61 16	93 576 664	.3072758	.6857328
7	13 46 89	49 430 863	.1572441	.1595988	455	20 70 25	94 196 375	21.3307290	7.6913717
8	13 54 24	49 836 032	.1833261	.1660957	6	20 79 36	94 818 816	.3541565	.6970023
9	13 61 61	50 243 409	.2093727	.1725809	7	20 88 49	95 443 993	.3775583	.7026246
370	13 69 00	50 653 000	19.2353841	7.1790544	8	20 97 64	96 071 912	.4009346	.7082388
1	13 76 41	51 064 811	.2613603	.1855162	9	21 06 81	96 702 579	.4242853	.7138448
2	13 83 84	51 478 848	.2873015	.1919663	460	21 16 00	97 336 000	21.4476106	7.7194426
3	13 91 29	51 895 117	.3132079	.1984050	1	21 25 21	97 972 181	.4709106	.7250325
4	13 98 76	52 313 624	.3390796	.2048322	2	21 34 44	98 611 128	.4941853	.7306141
375	14 06 25	52 734 375	19.3649167	7.2112479	3	21 43 69	99 252 847	.5174348	.7361877
6	14 13 76	53 157 376	.3907194	.2176522	4	21 52 96	99 897 344	.5406592	.7417532
7	14 21 29	53 582 633	.4164878	.2240450	465	21 62 25	100 544 625	21.5638587	7.7473109
8	14 28 84	54 010 152	.4422221	.2304268	6	21 71 56	101 194 696	.5870331	.7528606
9	14 36 41	54 439 939	.4679223	.2367972	7	21 80 89	101 847 563	.6101828	.7584023
380	14 44 00	54 872 000	19.4935887	7.2431565	8	21 90 24	102 503 232	.6333077	.7639361
1	14 51 61	55 306 341	.5192213	.2495045	9	21 99 61	103 161 709	.6564078	.7694620
2	14 59 24	55 742 968	.5448203	.2558415	470	22 09 00	103 823 000	21.6794834	7.7749801
3	14 66 89	56 181 887	.5703858	.2621675	1	22 18 41	104 487 111	.7025344	.7804904
4	14 74 56	56 623 104	.5959179	.2684824	2	22 27 84	105 154 048	.7255610	.7859928
385	14 82 25	57 066 625	19.6214169	7.2747864	3	22 37 29	105 823 817	.7485632	.7914875
6	14 89 96	57 512 456	.6468827	.2810794	4	22 46 76	106 496 424	.7715411	.7969745
7	14 97 69	57 960 603	.6723156	.2873617	475	22 56 25	107 171 875	21.7944947	7.8024538
8	15 05 44	58 411 072	.6977156	.2936330	6	22 65 76	107 850 176	.8174242	.8079253
9	15 13 21	58 863 869	.7230829	.2998936	7	22 75 29	108 531 333	.8403297	.8133892
390	15 21 00	59 319 000	19.7484177	7.3061436	8	22 84 84	109 215 352	.8632111	.8188456
1	15 28 81	59 776 471	.7737199	.3123828	9	22 94 41	109 902 239	.8860686	.8242942
2	15 36 64	60 236 288	.7989899	.3186114	480	23 04 00	110 592 000	21.9089023	7.8297353
3	15 44 49	60 698 457	.8242276	.3248295	1	23 13 61	111 284 641	.9317122	.8351688
4	15 52 36	61 162 984	.8494332	.3310369	2	23 23 24	111 980 168	.9544984	.8405949
395	15 60 25	61 629 875	19.8746069	7.3372339	3	23 32 89	112 678 587	.9772610	.8460134
6	15 68 16	62 099 136	.8997487	.3434205	4	23 42 56	113 379 904	22.0000000	.8514244
7	15 76 09	62 570 773	.9248588	.3495966	485	23 52 25	114 084 125	22.0227155	7.8568281
8	15 84 04	63 044 792	.9499373	.3557624	6	23 61 96	114 791 256	.0454077	.8622242
9	15 92 01	63 521 199	.9749844	.3619178	7	23 71 69	115 501 303	.0680765	.8676130
400	16 00 00	64 000 000	20.0000000	7.3680630	8	23 81 44	116 214 272	.0907220	.8729944
1	16 08 01	64 481 201	.0249844	.3741979	9	23 91 21	116 930 169	.1133444	.8783684
2	16 16 04	64 964 808	.0499377	.3803227	490	24 01 00	117 649 000	22.1359436	7.8837352
3	16 24 09	65 450 827	.0748599	.3864373	1	24 10 81	118 370 771	.1585198	.8890946
4	16 32 16	65 939 264	.0997512	.3925418	2	24 20 64	119 095 488	.1810730	.8944468
405	16 40 25	66 430 125	20.1246118	7.3986363	3	24 30 49	119 823 157	.2036033	.8997917
6	16 48 36	66 923 416	.1494417	.4047206	4	24 40 36	120 553 784	.2261108	.9051294
7	16 56 49	67 419 143	.1742410	.4107950	495	24 50 25	121 287 375	22.2485955	7.9104599
8	16 64 64	67 917 312	.1990099	.4168595	6	24 60 16	122 023 936	.2710575	.9157832
9	16 72 81	68 417 929	.2237484	.4229142	7	24 70 09	122 763 473	.2934968	.9210994
410	16 81 00	68 921 000	20.2484567	7.4289589	8	24 80 04	123 505 992	.3159136	.9264085
11	16 89 21	69 426 531	.2731349	.4349938	9	24 90 01	124 251 499	.3383079	.9317104
12	16 97 44	69 934 528	.2977831	.4410189	500	25 00 00	125 000 000	22.3606798	7.9370053
13	17 05 69	70 444 997	.3224014	.4470342	1	25 10 01	125 751 501	.3830293	.9422931
14	17 13 96	70 957 944	.3469899	.4530399	2	25 20 04	126 506 008	.4053565	.9475739
415	17 22 25	71 473 375	20.3715488	7.4590359	3	25 30 09	127 263 527	.4276615	.9528477
16	17 30 56	71 991 296	.3960781	.4650223	4	25 40 16	128 024 064	.4499443	.9581144
17	17 38 89	72 511 713	.4205779	.4709991	505	25 50 25	128 787 625	22.4722051	7.9633743
18	17 47 24	73 034 632	.4450483	.4769664	6	25 60 36	129 554 216	.4944438	.9686271
19	17 55 61	73 560 059	.4694895	.4829242	7	25 70 49	130 323 843	.5166605	.9738731
420	17 64 00	74 088 000	20.4939015	7.4888724	8	25 80 64	131 096 512	.5388553	.9791122
1	17 72 41	74 618 461	.5182845	.4948113	9	25 90 81	131 872 229	.5610283	.9843444
2	17 80 84	75 151 448	.5426386	.5007406	510	26 01 00	132 651 000	22.5831796	7.9895697
3	17 89 29	75 686 967	.5669638	.5066607	11	26 11 21	133 432 831	.6053091	.9947883
4	17 97 76	76 225 024	.5912603	.5125715	12	26 21 44	134 217 728	.6274170	8.0000000
425	18 06 25	76 765 625	20.6155281	7.5184730	13	26 31 69	135 005 697	.6495033	.0052049
6	18 14 76	77 308 776	.6397674	.5243652	14	26 41 96	135 796 744	.6715681	.0104032
7	18 23 29	77 854 483	.6639783	.5302482	515	26 52 25	136 590 875	22.6936114	8.0155946
8	18 31 84	78 402 752	.6881609	.5361221	16	26 62 56	137 388 096	.7156334	.0207794
9	18 40 41	78 953 589	.7123152	.5419867	17	26 72 89	138 188 413	.7376340	.0259574
430	18 49 00	79 507 000	20.7364414	7.5478423	18	26 83 24	138 991 832	.7596134	.0311287
1	18 57 61	80 062 991	.7605395	.5536888	19	26 93 61	139 798 359	.7815715	.0362935
2	18 66 24	80 621 568	.7846097	.5595263	520	27 04 00	140 608 000	22.8035085	8.0414515
3	18 74 89	81 182 737	.8086520	.5653548	1	27 14 41	141 420 761	.8254244	.0466030
4	18 83 56	81 746 504	.8326667	.5711743	2	27 24 84	142 236 648	.8473193	.0517479
435	18 92 25	82 312 875	20.8566536	7.5769849	3	27 35 29	143 055 667	.8691933	.0568862
6	19 00 96	82 881 856	.8806130	.5827865	4	27 45 76	143 877 824	.8910463	.0620180

No.	Square	Cube	Sq. Rt.	Cu. Rt.	No.	Square	Cube	Sq. Rt.	Cu. Rt.
4	27 45 76	143 877 824	.8910463	.0620180	12	37 45 44	229 220 928	.7386338	.4901848
525	27 56 25	144 703 125	22.9128785	8.0671432	13	37 57 69	230 346 397	.7588368	.4948065
6	27 66 76	145 531 576	.9346899	.0722620	14	37 69 96	231 475 544	.7790234	.4994233
7	27 77 29	146 363 183	.9564806	.0773743	615	37 82 25	232 608 375	24.7991935	8.5040350
8	27 87 84	147 197 952	.9782506	.0824800	16	37 94 56	233 744 896	.8193473	.5086417
9	27 98 41	148 035 889	23.0000000	.0875794	17	38 06 89	234 885 113	.8394847	.5132435
530	28 09 00	148 877 000	23.0217289	8.0926723	18	38 19 24	236 029 032	.8596058	.5178403
1	28 19 61	149 721 291	.0434372	.0977589	19	38 31 61	237 176 659	.8797106	.5224321
2	28 30 24	150 568 768	.0651252	.1028390	620	38 44 00	238 328 000	24.8997992	8.5270189
3	28 40 89	151 419 437	.0867928	.1079128	1	38 56 41	239 483 061	.9198716	.5316009
4	28 51 56	152 273 304	.1084400	.1129803	2	38 68 84	240 641 848	.9399278	.5361780
535	28 62 25	153 130 375	23.1300670	8.1180414	3	38 81 29	241 804 367	.9599679	.5407501
6	28 72 96	153 990 656	.1516738	.1230962	4	38 93 76	242 970 624	.9799920	.5453173
7	28 83 69	154 854 153	.1732605	.1281447	625	39 06 25	244 140 625	25.0000000	8.5498797
8	28 94 44	155 720 872	.1948270	.1331870	6	39 18 76	245 314 376	.0199920	.5544372
9	29 05 21	156 590 819	.2163735	.1382230	7	39 31 29	246 491 883	.0399681	.5589899
540	29 16 00	157 464 000	23.2379001	8.1432529	8	39 43 84	247 673 152	.0599282	.5635377
1	29 26 81	158 340 421	.2594067	.1482765	9	39 56 41	248 858 189	.0798724	.5680807
2	29 37 64	159 220 088	.2808935	.1532939	630	39 69 00	250 047 000	25.0998008	8.5726189
3	29 48 49	160 103 007	.3023604	.1583051	1	39 81 61	251 239 591	.1197134	.5771523
4	29 59 36	160 989 184	.3238076	.1633102	2	39 94 24	252 435 968	.1396102	.5816809
545	29 70 25	161 878 625	23.3452351	8.1683092	3	40 06 89	253 636 137	.1594913	.5862047
6	29 81 16	162 771 336	.3666429	.1733020	4	40 19 56	254 840 104	.1793566	.5907238
7	29 92 09	163 667 323	.3880311	.1782888	635	40 32 25	256 047 875	25.1992063	8.5952380
8	30 03 04	164 566 592	.4093998	.1832695	6	40 44 96	257 259 456	.2190404	.5997476
9	30 14 01	165 469 149	.4307490	.1882441	7	40 57 69	258 474 853	.2388589	.6042525
550	30 25 00	166 375 000	23.4520788	8.1932127	8	40 70 44	259 694 072	.2586619	.6087526
1	30 36 01	167 284 151	.4733892	.1981753	9	40 83 21	260 917 119	.2784493	.6132480
2	30 47 04	168 196 608	.4946802	.2031319	640	40 96 00	262 144 000	25.2982213	8.6177388
3	30 58 09	169 112 377	.5159520	.2080825	1	41 08 81	263 374 721	.3179778	.6222248
4	30 69 16	170 031 464	.5372046	.2130271	2	41 21 64	264 609 288	.3377189	.6267063
555	30 80 25	170 953 875	23.5584380	8.2179657	3	41 34 49	265 847 707	.3574447	.6311830
6	30 91 36	171 879 616	.5796522	.2228985	4	41 47 36	267 089 984	.3771551	.6356551
7	31 02 49	172 808 693	.6008474	.2278254	645	41 60 25	268 336 125	25.3968502	8.6401226
8	31 13 64	173 741 112	.6220236	.2327463	6	41 73 16	269 586 136	.4165301	.6445855
9	31 24 81	174 676 879	.6431808	.2376614	7	41 86 09	270 840 023	.4361947	.6490437
560	31 36 00	175 616 000	23.6643191	8.2425706	8	41 99 04	272 097 792	.4558441	.6534974
1	31 47 21	176 558 481	.6854386	.2474740	9	42 12 01	273 359 449	.4754784	.6579465
2	31 58 44	177 504 328	.7065392	.2523715	650	42 25 00	274 625 000	25.4950976	8.6623911
3	31 69 69	178 453 547	.7276210	.2572633	1	42 38 01	275 894 451	.5147016	.6668310
4	31 80 96	179 406 144	.7486842	.2621492	2	42 51 04	277 167 808	.5342907	.6712665
565	31 92 25	180 362 125	23.7697286	8.2670294	3	42 64 09	278 445 077	.5538647	.6756974
6	32 03 56	181 321 496	.7907545	.2719039	4	42 77 16	279 726 264	.5734237	.6801237
7	32 14 89	182 284 263	.8117618	.2767726	655	42 90 25	281 011 375	25.5929678	8.6845456
8	32 26 24	183 250 432	.8327506	.2816355	6	43 03 36	282 300 416	.6124969	.6889630
9	32 37 61	184 220 009	.8537209	.2864928	7	43 16 49	283 593 393	.6320112	.6933759
570	32 49 00	185 193 000	23.8746728	8.2913444	8	43 29 64	284 890 312	.6515107	.6977843
1	32 60 41	186 169 411	.8956063	.2961903	9	43 42 81	286 191 179	.6709953	.7021882
2	32 71 84	187 149 248	.9165215	.3010304	660	43 56 00	287 496 000	25.6904652	8.7065877
3	32 83 29	188 132 517	.9374184	.3058651	1	43 69 21	288 804 781	.7099203	.7109827
4	32 94 76	189 119 224	.9582971	.3106941	2	43 82 44	290 117 528	.7293607	.7153734
575	33 06 25	190 109 375	23.9791576	8.3155175	3	43 95 69	291 434 247	.7487864	.7197596
6	33 17 76	191 102 976	24.0000000	.3203353	4	44 08 96	292 754 944	.7681975	.7241414
7	33 29 29	192 100 033	.0208243	.3251475	665	44 22 25	294 079 625	25.7875939	8.7285187
8	33 40 84	193 100 552	.0416306	.3299542	6	44 35 56	295 408 296	.8069758	.7328918
9	33 52 41	194 104 539	.0624188	.3347553	7	44 48 89	296 740 963	.8263431	.7372604
580	33 64 00	195 112 000	24.0831891	8.3395509	8	44 62 24	298 077 632	.8456960	.7416246
1	33 75 61	196 122 941	.1039416	.3443410	9	44 75 61	299 418 309	.8650343	.7459846
2	33 87 24	197 137 368	.1246762	.3491256	670	44 89 00	300 763 000	25.8843582	8.7503401
3	33 98 89	198 155 287	.1453929	.3539047	1	45 02 41	302 111 711	.9036677	.7546913
4	34 10 56	199 176 704	.1660919	.3586784	2	45 15 84	303 464 448	.9229628	.7590383
585	34 22 25	200 201 625	24.1867732	8.3634466	3	45 29 29	304 821 217	.9422435	.7633809
6	34 33 96	201 230 056	.2074369	.3682095	4	45 42 76	306 182 024	.9615100	.7677192
7	34 45 69	202 262 003	.2280829	.3729668	675	45 56 25	307 546 875	25.9807621	8.7720532
8	34 57 44	203 297 472	.2487113	.3777188	6	45 69 76	308 915 776	26.0000000	.7763830
9	34 69 21	204 336 469	.2693222	.3824653	7	45 83 29	310 288 733	.0192237	.7807084
590	34 81 00	205 379 000	24.2899156	8.3872065	8	45 96 84	311 665 752	.0384331	.7850296
1	34 92 81	206 425 071	.3104916	.3919423	9	46 10 41	313 046 839	.0576284	.7893466
2	35 04 64	207 474 688	.3310501	.3966729	680	46 24 00	314 432 000	26.0768096	8.7936593
3	35 16 49	208 527 857	.3515913	.4013981	1	46 37 61	315 821 241	.0959767	.7979679
4	35 28 36	209 584 584	.3721152	.4061180	2	46 51 24	317 214 568	.1151297	.8022721
595	35 40 25	210 644 875	24.3926218	8.4108326	3	46 64 89	318 611 987	.1342687	.8065722
6	35 52 16	211 708 736	.4131112	.4155419	4	46 78 56	320 013 504	.1533937	.8108681
7	35 64 09	212 776 173	.4335834	.4202460	685	46 92 25	321 419 125	26.1725047	8.8151598
8	35 76 04	213 847 192	.4540385	.4249448	6	47 05 96	322 828 856	.1916017	.8194474
9	35 88 01	214 921 799	.4744765	.4296383	7	47 19 69	324 242 703	.2106848	.8237307
600	36 00 00	216 000 000	24.4948974	8.4343267	8	47 33 44	325 660 672	.2297541	.8280099
1	36 12 01	217 081 801	.5153013	.4390098	9	47 47 21	327 082 769	.2488095	.8322850
2	36 24 04	218 167 208	.5356883	.4436877	690	47 61 00	328 509 000	26.2678511	8.8365559
3	36 36 09	219 256 227	.5560583	.4483605	1	47 74 81	329 939 371	.2868789	.8408227
4	36 48 16	220 348 864	.5764115	.4530281	2	47 88 64	331 373 888	.3058929	.8450854
605	36 60 25	221 445 125	24.5967478	8.4576906	3	48 02 49	332 812 557	.3248932	.8493440
6	36 72 36	222 545 016	.6170673	.4623479	4	48 16 36	334 255 384	.3438797	.8535985
7	36 84 49	223 648 543	.6373700	.4670000	695	48 30 25	335 702 375	26.3628527	8.8578489
8	36 96 64	224 755 712	.6576560	.4716471	6	48 44 16	337 153 536	.3818119	.8620952
9	37 08 81	225 866 529	.6779254	.4762892	7	48 58 09	338 608 873	.4007576	.8663375
610	37 21 00	226 981 000	24.6981781	8.4809261	8	48 72 04	340 068 392	.4196896	.8705757
11	37 33 21	228 099 131	.7184142	.4855579	9	48 86 01	341 532 099	.4386081	.8748099
12	37 45 44	229 220 928	.7386338	.4901848	700	49 00 00	343 000 000	26.4575131	8.8790400

No.	Square	Cube	Sq. Rt.	Cu. Rt.	No.	Square	Cube	Sq. Rt.	Cu. Rt.
700	49 00 00	343 000 000	26.4575131	8.8790400	8	62 09 44	489 303 872	.0713377	.2365277
1	49 14 01	344 472 101	.4764046	.8832661	9	62 25 21	491 169 069	.0891438	.2404333
2	49 28 04	345 948 408	.4952826	.8874882	790	62 41 00	493 039 000	28.1069386	9.2443355
3	49 42 09	347 428 927	.5141472	.8917063	1	62 56 81	494 913 671	.1247222	.2482344
4	49 56 16	348 913 664	.5329983	.8959204	2	62 72 64	496 793 088	.1424946	.2521300
705	49 70 25	350 402 625	26.5518361	8.9001304	3	62 88 49	498 677 257	.1602557	.2560224
6	49 84 36	351 895 816	.5706605	.9043366	4	63 04 36	500 566 184	.1780056	.2599114
7	49 98 49	353 393 243	.5894716	.9085387	795	63 20 25	502 459 875	28.1957444	9.2637973
8	50 12 64	354 894 912	.6082694	.9127369	6	63 36 16	504 358 336	.2134720	.2676798
9	50 26 81	356 400 829	.6270539	.9169311	7	63 52 09	506 261 573	.2311884	.2715592
710	50 41 00	357 911 000	26.6458252	8.9211214	8	63 68 04	508 169 592	.2488938	.2754352
11	50 55 21	359 425 431	.6645833	.9253078	9	63 84 01	510 082 399	.2665881	.2793081
12	50 69 44	360 944 128	.6833281	.9294902	800	64 00 00	512 000 000	28.2842712	9.2831777
13	50 83 69	362 467 097	.7020598	.9336687	1	64 16 01	513 922 401	.3019434	.2870440
14	50 97 96	363 994 344	.7207784	.9378433	2	64 32 04	515 849 608	.3196045	.2909072
715	51 12 25	365 525 875	26.7394839	8.9420140	3	64 48 09	517 781 627	.3372546	.2947671
16	51 26 56	367 061 696	.7581763	.9461809	4	64 64 16	519 718 464	.3548938	.2986239
17	51 40 89	368 601 813	.7768557	.9503438	805	64 80 25	521 660 125	28.3725219	9.3024775
18	51 55 24	370 146 232	.7955220	.9545029	6	64 96 36	523 606 616	.3901391	.3063278
19	51 69 61	371 694 959	.8141754	.9586581	7	65 12 49	525 557 943	.4077454	.3101750
720	51 84 00	373 248 000	26.8328157	8.9628095	8	65 28 64	527 514 112	.4253408	.3140190
1	51 98 41	374 805 361	.8514432	.9669570	9	65 44 81	529 475 129	.4429253	.3178599
2	52 12 84	376 367 048	.8700577	.9711007	810	65 61 00	531 441 000	28.4604989	9.3216975
3	52 27 29	377 933 067	.8886593	.9752406	11	65 77 21	533 411 731	.4780617	.3255320
4	52 41 76	379 503 424	.9072481	.9793766	12	65 93 44	535 387 328	.4956137	.3293634
725	52 56 25	381 078 125	26.9258240	8.9835089	13	66 09 69	537 367 797	.5131549	.3331916
6	52 70 76	382 657 176	.9443872	.9876373	14	66 25 96	539 353 144	.5306852	.3370167
7	52 85 29	384 240 583	.9629375	.9917620	815	66 42 25	541 343 375	28.5482048	9.3408386
8	52 99 84	385 828 352	.9814751	.9958829	16	66 58 56	543 338 496	.5657137	.3446575
9	53 14 41	387 420 489	27.0000000	9.0000000	17	66 74 89	545 338 513	.5832119	.3484731
730	53 29 00	389 017 000	27.0185122	9.0041134	18	66 91 24	547 343 432	.6006993	.3522857
1	53 43 61	390 617 891	.0370117	.0082229	19	67 07 61	549 353 259	.6181760	.3560952
2	53 58 24	392 223 168	.0554985	.0123288	820	67 24 00	551 368 000	28.6356421	9.3599016
3	53 72 89	393 832 837	.0739727	.0164309	1	67 40 41	553 387 661	.6530976	.3637049
4	53 87 56	395 446 904	.0924344	.0205293	2	67 56 84	555 412 248	.6705424	.3675051
735	54 02 25	397 065 375	27.1108834	9.0246239	3	67 73 29	557 441 767	.6879766	.3713022
6	54 16 96	398 688 256	.1293199	.0287149	4	67 89 76	559 476 224	.7054002	.3750963
7	54 31 69	400 315 553	.1477439	.0328021	825	68 06 25	561 515 625	28.7228132	9.3788873
8	54 46 44	401 947 272	.1661554	.0368857	6	68 22 76	563 559 976	.7402157	.3826752
9	54 61 21	403 583 419	.1845544	.0409655	7	68 39 29	565 609 283	.7576077	.3864600
740	54 76 00	405 224 000	27.2029410	9.0450417	8	68 55 84	567 663 552	.7749891	.3902419
1	54 90 81	406 869 021	.2213152	.0491142	9	68 72 41	569 722 789	.7923601	.3940206
2	55 05 64	408 518 488	.2396769	.0531831	830	68 89 00	571 787 000	28.8097206	9.3977964
3	55 20 49	410 172 407	.2580263	.0572482	1	69 05 61	573 856 191	.8270706	.4015691
4	55 35 36	411 830 784	.2763634	.0613098	2	69 22 24	575 930 368	.8444102	.4053387
745	55 50 25	413 493 625	27.2946881	9.0653677	3	69 38 89	578 009 537	.8617394	.4091054
6	55 65 16	415 160 936	.3130006	.0694220	4	69 55 56	580 093 704	.8790582	.4128690
7	55 80 09	416 832 723	.3313007	.0734726	835	69 72 25	582 182 875	28.8963666	9.4166297
8	55 95 04	418 508 992	.3495887	.0775197	6	69 88 96	584 277 056	.9136646	.4203873
9	56 10 01	420 189 749	.3678644	.0815631	7	70 05 69	586 376 253	.9309523	.4241420
750	56 25 00	421 875 000	27.3861279	9.0856030	8	70 22 44	588 480 472	.9482297	.4278936
1	56 40 01	423 564 751	.4043792	.0896392	9	70 39 21	590 589 719	.9654967	.4316423
2	56 55 04	425 259 008	.4226184	.0936719	840	70 56 00	592 704 000	28.9827535	9.4353880
3	56 70 09	426 957 777	.4408455	.0977010	1	70 72 81	594 823 321	29.0000000	.4391307
4	56 85 16	428 661 064	.4590604	.1017265	2	70 89 64	596 947 688	.0172363	.4428704
755	57 00 25	430 368 875	27.4772633	9.1057485	3	71 06 49	599 077 107	.0344623	.4466072
6	57 15 36	432 081 216	.4954542	.1097669	4	71 23 36	601 211 584	.0516781	.4503410
7	57 30 49	433 798 093	.5136330	.1137818	845	71 40 25	603 351 125	29.0688837	9.4540719
8	57 45 64	435 519 512	.5317998	.1177931	6	71 57 16	605 495 736	.0860791	.4577999
9	57 60 81	437 245 479	.5499546	.1218010	7	71 74 09	607 645 423	.1032644	.4615249
760	57 76 00	438 976 000	27.5680975	9.1258053	8	71 91 04	609 800 192	.1204396	.4652470
1	57 91 21	440 711 081	.5862284	.1298061	9	72 08 01	611 960 049	.1376046	.4689661
2	58 06 44	442 450 728	.6043475	.1338034	850	72 25 00	614 125 000	29.1547595	9.4726824
3	58 21 69	444 194 947	.6224546	.1377971	1	72 42 01	616 295 051	.1719043	.4763957
4	58 36 96	445 943 744	.6405499	.1417874	2	72 59 04	618 470 208	.1890390	.4801061
765	58 52 25	447 697 125	27.6586334	9.1457743	3	72 76 09	620 650 477	.2061637	.4838136
6	58 67 56	449 455 096	.6767050	.1497576	4	72 93 16	622 835 864	.2232784	.4875182
7	58 82 89	451 217 663	.6947648	.1537375	855	73 10 25	625 026 375	29.2403830	9.4912200
8	58 98 24	452 984 832	.7128129	.1577139	6	73 27 36	627 222 016	.2574777	.4949188
9	59 13 61	454 756 609	.7308492	.1616869	7	73 44 49	629 422 793	.2745623	.4986147
770	59 29 00	456 533 000	27.7488739	9.1656565	8	73 61 64	631 628 712	.2916370	.5023078
1	59 44 41	458 314 011	.7668868	.1696225	9	73 78 81	633 839 779	.3087018	.5059980
2	59 59 84	460 099 648	.7848880	.1735852	860	73 96 00	636 056 000	29.3257566	9.5096854
3	59 75 29	461 889 917	.8028775	.1775445	1	74 13 21	638 277 381	.3428015	.5133699
4	59 90 76	463 684 824	.8208555	.1815003	2	74 30 44	640 503 928	.3598365	.5170515
775	60 06 25	465 484 375	27.8388218	9.1854527	3	74 47 69	642 735 647	.3768616	.5207303
6	60 21 76	467 288 576	.8567766	.1894018	4	74 64 96	644 972 544	.3938769	.5244063
7	60 37 29	469 097 433	.8747197	.1933474	865	74 82 25	647 214 625	29.4108823	9.5280794
8	60 52 84	470 910 952	.8926514	.1972897	6	74 99 56	649 461 896	.4278779	.5317497
9	60 68 41	472 729 139	.9105715	.2012286	7	75 16 89	651 714 363	.4448637	.5354172
780	60 84 00	474 552 000	27.9284801	9.2051641	8	75 34 24	653 972 032	.4618397	.5390818
1	60 99 61	476 379 541	.9463772	.2090962	9	75 51 61	656 234 909	.4788059	.5427437
2	61 15 24	478 211 768	.9642629	.2130250	870	75 69 00	658 503 000	29.4957624	9.5464027
3	61 30 89	480 048 687	.9821372	.2169505	1	75 86 41	660 776 311	.5127091	.5500589
4	61 46 56	481 890 304	28.0000000	.2208726	2	76 03 84	663 054 848	.5296461	.5537123
785	61 62 25	483 736 625	28.0178515	9.2247914	3	76 21 29	665 338 617	.5465734	.5573630
6	61 77 96	485 587 656	.0356915	.2287068	4	76 38 76	667 627 624	.5634910	.5610108
7	61 93 69	487 443 403	.0535203	.2326189	875	76 56 25	669 921 875	29.5803989	9.5646559
8	62 09 44	489 303 872	.0713377	.2365277	6	76 73 76	672 221 376	.5972972	.5682982

No.	Square	Cube	Sq. Rt.	Cu. Rt.	No.	Square	Cube	Sq. Rt.	Cu. Rt.
6	76 73 76	672 221 376	.5972972	.5682982	4	92 92 96	895 841 344	.0483494	.8785305
7	76 91 29	674 526 133	.6141858	.5719377	965	93 12 25	898 632 125	31.0644491	9.8819451
8	77 08 84	676 836 152	.6310648	.5755745	6	93 31 56	901 428 696	.0805405	.8853574
9	77 26 41	679 151 439	.6479342	.5792085	7	93 50 89	904 231 063	.0966236	.8887673
880	77 44 00	681 472 000	29.6647939	9.5828397	8	93 70 24	907 039 232	.1126984	.8921749
1	77 61 61	683 797 841	.6816442	.5864682	9	93 89 61	909 853 209	.1287648	.8955801
2	77 79 24	686 128 968	.6984848	.5900939	970	94 09 00	912 673 000	31.1448230	9.8989830
3	77 96 89	688 465 387	.7153159	.5937169	1	94 28 41	915 498 611	.1608729	.9023835
4	78 14 56	690 807 104	.7321375	.5973373	2	94 47 84	918 330 048	.1769145	.9057817
885	78 32 25	693 154 125	29.7489496	9.6009548	3	94 67 29	921 167 317	.1929479	.9091776
6	78 49 96	695 506 456	.7657521	.6045696	4	94 86 76	924 010 424	.2089731	.9125712
7	78 67 69	697 864 103	.7825452	.6081817	975	95 06 25	926 859 375	31.2249900	9.9159624
8	78 85 44	700 227 072	.7993289	.6117911	6	95 25 76	929 714 176	.2409987	.9193513
9	79 03 21	702 595 369	.8161030	.6153977	7	95 45 29	932 574 833	.2569992	.9227379
890	79 21 00	704 969 000	29.8328678	9.6190017	8	95 64 84	935 441 352	.2729915	.9261222
1	79 38 81	707 347 971	.8496231	.6226030	9	95 84 41	938 313 739	.2889757	.9295042
2	79 56 64	709 732 288	.8663690	.6262016	980	96 04 00	941 192 000	31.3049517	9.9328839
3	79 74 49	712 121 957	.8831056	.6297975	1	96 23 61	944 076 141	.3209195	.9362613
4	79 92 36	714 516 984	.8998328	.6333907	2	96 43 24	946 966 168	.3368792	.9396363
895	80 10 25	716 917 375	29.9165506	9.6369812	3	96 62 89	949 862 087	.3528308	.9430092
6	80 28 16	719 323 136	.9332591	.6405690	4	96 82 56	952 763 904	.3687743	.9463797
7	80 46 09	721 734 273	.9499583	.6441542	985	97 02 25	955 671 625	31.3847097	9.9497479
8	80 64 04	724 150 792	.9666481	.6477367	6	97 21 96	958 585 256	.4006369	.9531138
9	80 82 01	726 572 699	.9833287	.6513166	7	97 41 69	961 504 803	.4165561	.9564775
900	81 00 00	729 000 000	30.0000000	9.6548938	8	97 61 44	964 430 272	.4324673	.9598389
1	81 18 01	731 432 701	.0166620	.6584684	9	97 81 21	967 361 669	.4483704	.9631981
2	81 36 04	733 870 808	.0333148	.6620403	990	98 01 00	970 299 000	31.4642654	9.9665549
3	81 54 09	736 314 327	.0499584	.6656096	1	98 20 81	973 242 271	.4801525	.9699095
4	81 72 16	738 763 264	.0665928	.6691762	2	98 40 64	976 191 488	.4960315	.9732619
905	81 90 25	741 217 625	30.0832179	9.6727403	3	98 60 49	979 146 657	.5119025	.9766120
6	82 08 36	743 677 416	.0998339	.6763017	4	98 80 36	982 107 784	.5277655	.9799599
7	82 26 49	746 142 643	.1164407	.6798604	995	99 00 25	985 074 875	31.5436206	9.9833055
8	82 44 64	748 613 312	.1330383	.6834166	6	99 20 16	988 047 936	.5594677	.9866488
9	82 62 81	751 089 429	.1496269	.6869701	7	99 40 09	991 026 973	.5753068	.9899900
910	82 81 00	753 571 000	30.1662063	9.6905211	8	99 60 04	994 011 992	.5911380	.9933289
11	82 99 21	756 058 031	.1827765	.6940694	9	99 80 01	997 002 999	.6069613	.9966656
12	83 17 44	758 550 528	.1993377	.6976151	1000	1 00 00 00	1 000 000 000	31.6227766	10.0000000
13	83 35 69	761 048 497	.2158899	.7011583	1	1 00 20 01	1 003 003 001	.6385840	.0033322
14	83 53 96	763 551 944	.2324329	.7046989	2	1 00 40 04	1 006 012 008	.6543836	.0066622
915	83 72 25	766 060 875	30.2489669	9.7082369	3	1 00 60 09	1 009 027 027	.6701752	.0099899
16	83 90 56	768 575 296	.2654919	.7117723	4	1 00 80 16	1 012 048 064	.6859590	.0133155
17	84 08 89	771 095 213	.2820079	.7153051	1005	1 01 00 25	1 015 075 125	31.7017349	10.0166389
18	84 27 24	773 620 632	.2985148	.7188354	6	1 01 20 36	1 018 108 216	.7175030	.0199601
19	84 45 61	776 151 559	.3150128	.7223631	7	1 01 40 49	1 021 147 343	.7332633	.0232791
920	84 64 00	778 688 000	30.3315018	9.7258883	8	1 01 60 64	1 024 192 512	.7490157	.0265958
1	84 82 41	781 229 961	.3479818	.7294109	9	1 01 80 81	1 027 243 729	.7647603	.0299104
2	85 00 84	783 777 448	.3644529	.7329309	1010	1 02 01 00	1 030 301 000	31.7804972	10.0332228
3	85 19 29	786 330 467	.3809151	.7364484	11	1 02 21 21	1 033 364 331	.7962262	.0365330
4	85 37 76	788 889 024	.3973683	.7399634	12	1 02 41 44	1 036 433 728	.8119474	.0398410
925	85 56 25	791 453 125	30.4138127	9.7434758	13	1 02 61 69	1 039 509 197	.8276609	.0431469
6	85 74 76	794 022 776	.4302481	.7469857	14	1 02 81 96	1 042 590 744	.8433666	.0464506
7	85 93 29	796 597 983	.4466747	.7504930	1015	1 03 02 25	1 045 678 375	31.8590646	10.0497521
8	86 11 84	799 178 752	.4630924	.7539979	16	1 03 22 56	1 048 772 096	.8747549	.0530514
9	86 30 41	801 765 089	.4795013	.7575002	17	1 03 42 89	1 051 871 913	.8904374	.0563485
930	86 49 00	804 357 000	30.4959014	9.7610001	18	1 03 63 24	1 054 977 832	.9061123	.0596435
1	86 67 61	806 954 491	.5122926	.7644974	19	1 03 83 61	1 058 089 859	.9217794	.0629364
2	86 86 24	809 557 568	.5286750	.7679922	1020	1 04 04 00	1 061 208 000	31.9374388	10.0662271
3	87 04 89	812 166 237	.5450487	.7714845	21	1 04 24 41	1 064 332 261	.9530906	.0695156
4	87 23 56	814 780 504	.5614136	.7749743	22	1 04 44 84	1 067 462 648	.9687347	.0728020
935	87 42 25	817 400 375	30.5777697	9.7784616	23	1 04 65 29	1 070 599 167	.9843712	.0760863
6	87 60 96	820 025 856	.5941171	.7819466	24	1 04 85 76	1 073 741 824	32.0000000	.0793684
7	87 79 69	822 656 953	.6104557	.7854288	1025	1 05 06 25	1 076 890 625	32.0156212	10.0826484
8	87 98 44	825 293 672	.6267857	.7889087	26	1 05 26 76	1 080 045 576	.0312348	.0859262
9	88 17 21	827 936 019	.6431069	.7923861	27	1 05 47 29	1 083 206 683	.0468407	.0892019
940	88 36 00	830 584 000	30.6594194	9.7958611	28	1 05 67 84	1 086 373 952	.0624391	.0924755
1	88 54 81	833 237 621	.6757233	.7993336	29	1 05 88 41	1 089 547 389	.0780298	.0957469
2	88 73 64	835 896 888	.6920185	.8028036	1030	1 06 09 00	1 092 727 000	32.0936131	10.0990163
3	88 92 49	838 561 807	.7083051	.8062711	31	1 06 29 61	1 095 912 791	.1091887	.1022835
4	89 11 36	841 232 384	.7245830	.8097363	32	1 06 50 24	1 099 104 768	.1247568	.1055487
945	89 30 25	843 908 625	30.7408523	9.8131989	33	1 06 70 89	1 102 302 937	.1403173	.1088117
6	89 49 16	846 590 536	.7571130	.8166591	34	1 06 91 56	1 105 507 304	.1558704	.1120726
7	89 68 09	849 278 123	.7733651	.8201169	1035	1 07 12 25	1 108 717 875	32.1714159	10.1153314
8	89 87 04	851 971 392	.7896086	.8235723	36	1 07 32 96	1 111 934 656	.1869539	.1185882
9	90 06 01	854 670 349	.8058436	.8270252	37	1 07 53 69	1 115 157 653	.2024844	.1218428
950	90 25 00	857 375 000	30.8220700	9.8304757	38	1 07 74 44	1 118 386 872	.2180074	.1250953
1	90 44 01	860 085 351	.8382879	8339238	39	1 07 95 21	1 121 622 319	2335229	.1283457
2	90 63 04	862 801 408	.8544972	8373695	1040	1 08 16 00	1 124 864 000	32.2490310	10.1315941
3	90 82 09	865 523 177	.8706981	8408127	41	1 08 36 81	1 128 111 921	.2645316	.1348403
4	91 01 16	868 250 664	.8868904	.8442536	42	1 08 57 64	1 131 366 088	.2800248	.1380845
955	91 20 25	870 983 875	30.9030743	9.8476920	43	1 08 78 49	1 134 626 507	.2955105	.1413266
6	91 39 36	873 722 816	.9192497	.8511280	44	1 08 99 36	1 137 893 184	.3109888	.1445666
7	91 58 49	876 467 493	.9354166	.8545617	1045	1 09 20 25	1 141 166 125	32.3264598	10.1478047
8	91 77 64	879 217 912	.9515751	.8579929	46	1 09 41 16	1 144 445 336	.3419233	.1510406
9	91 96 81	881 974 079	.9677251	.8614218	47	1 09 62 09	1 147 730 823	.3573794	.1542744
960	92 16 00	884 736 000	30.9838668	9.8648483	48	1 09 83 04	1 151 022 592	.3728281	.1575062
1	92 35 21	887 503 681	31.0000000	.8682724	49	1 10 04 01	1 154 320 649	.3882695	.1607359
2	92 54 44	890 277 128	.0161248	.8716941	1050	1 10 25 00	1 157 625 000	32.4037035	10.1639636
3	92 73 69	893 056 347	.0322413	.8751135	51	1 10 46 01	1 160 935 651	.4191301	.1671893
4	92 92 96	895 841 344	.0483494	.8785305	52	1 10 67 04	1 164 252 608	.4345495	.1704129

No.	Square	Cube	Sq. Rt.	Cu. Rt.	No.	Square	Cube	Sq. Rt.	Cu. Rt.
52	1 10 67 04	1 164 252 608	.4345495	.1704129	1140	1 29 96 00	1 481 544 000	33.7638860	10.4464393
53	1 10 88 09	1 167 575 877	.4499615	.1736344	41	1 30 18 81	1 485 446 221	.7786915	.4494929
54	1 11 09 16	1 170 905 464	.4653662	.1768539	42	1 30 41 64	1 489 355 288	.7934905	.4525448
1055	1 11 30 25	1 174 241 375	32.4807635	10.1800714	43	1 30 64 49	1 493 271 207	.8082830	.4555948
56	1 11 51 36	1 177 583 616	.4961536	.1832868	44	1 30 87 36	1 497 193 984	.8230691	.4586431
57	1 11 72 49	1 180 932 193	.5115364	.1865002	1145	1 31 10 25	1 501 123 625	33.8378486	10.4616896
58	1 11 93 64	1 184 287 112	.5269119	.1897116	46	1 31 33 16	1 505 060 136	.8526218	.4647343
59	1 12 14 81	1 187 648 379	.5422802	.1929209	47	1 31 56 09	1 509 003 523	.8673884	.4677773
1060	1 12 36 00	1 191 016 000	32.5576412	10.1961283	48	1 31 79 04	1 512 953 792	.8821487	.4708185
61	1 12 57 21	1 194 389 981	.5729949	.1993336	49	1 32 02 01	1 516 910 949	.8969025	.4738579
62	1 12 78 44	1 197 770 328	.5883415	.2025369	1150	1 32 25 00	1 520 875 000	33.9116499	10.4768955
63	1 12 99 69	1 201 157 047	.6036807	.2057382	51	1 32 48 01	1 524 845 951	.9263909	.4799314
64	1 13 20 96	1 204 550 144	.6190129	.2089375	52	1 32 71 04	1 528 823 808	.9411255	.4829656
1065	1 13 42 25	1 207 949 625	32.6343377	10.2121347	53	1 32 94 09	1 532 808 577	.9558537	.4859980
66	1 13 63 56	1 211 355 496	.6496554	.2153300	54	1 33 17 16	1 536 800 264	.9705755	.4890286
67	1 13 84 89	1 214 767 763	.6649659	.2185233	1155	1 33 40 25	1 540 798 875	33.9852910	10.4920575
68	1 14 06 24	1 218 186 432	.6802693	.2217146	56	1 33 63 36	1 544 804 416	34.0000000	.4950847
69	1 14 27 61	1 221 611 509	.6955654	.2249039	57	1 33 86 49	1 548 816 893	.0147027	.4981101
1070	1 14 49 00	1 225 043 000	32.7108544	10.2280912	58	1 34 09 64	1 552 836 312	.0293990	.5011337
71	1 14 70 41	1 228 480 911	.7261363	.2312766	59	1 34 32 81	1 556 862 679	.0440890	.5041556
72	1 14 91 84	1 231 925 248	.7414111	.2344599	1160	1 34 56 00	1 560 896 000	34.0587727	10.5071757
73	1 15 13 29	1 235 376 017	.7566787	.2376413	61	1 34 79 21	1 564 936 281	.0734501	.5101942
74	1 15 34 76	1 238 833 224	.7719392	.2408207	62	1 35 02 44	1 568 983 528	.0881211	.5132109
1075	1 15 56 25	1 242 296 875	32.7871926	10.2439981	63	1 35 25 69	1 573 037 747	.1027858	.5162259
76	1 15 77 76	1 245 766 976	.8024389	.2471735	64	1 35 48 96	1 577 098 944	.1174442	.5192391
77	1 15 99 29	1 249 243 533	.8176782	.2503470	1165	1 35 72 25	1 581 167 125	34.1320963	10.5222506
78	1 16 20 84	1 252 726 552	.8329103	.2535186	66	1 35 95 56	1 585 242 296	.1467422	.5252604
79	1 16 42 41	1 256 216 039	.8481354	.2566881	67	1 36 18 89	1 589 324 463	.1613817	.5282685
1080	1 16 64 00	1 259 712 000	32.8633535	10.2598557	68	1 36 42 24	1 593 413 632	.1760150	.5312749
81	1 16 85 61	1 263 214 441	.8785644	.2630213	69	1 36 65 61	1 597 509 809	.1906420	.5342795
82	1 17 07 24	1 266 723 368	.8937684	.2661850	1170	1 36 89 00	1 601 613 000	34.2052627	10.5372825
83	1 17 28 89	1 270 238 787	.9089653	.2693467	71	1 37 12 41	1 605 723 211	.2198773	.5402837
84	1 17 50 56	1 273 760 704	.9241553	.2725065	72	1 37 35 84	1 609 840 448	.2344855	.5432832
1085	1 17 72 25	1 277 289 125	32.9393382	10.2756644	73	1 37 59 29	1 613 964 717	.2490875	.5462810
86	1 17 93 96	1 280 824 056	.9545141	.2788203	74	1 37 82 76	1 618 096 024	.2636834	.5492771
87	1 18 15 69	1 284 365 503	.9696830	.2819743	1175	1 38 06 25	1 622 234 375	34.2782730	10.5522715
88	1 18 37 44	1 287 913 472	.9848450	.2851264	76	1 38 29 76	1 626 379 776	.2928564	.5552642
89	1 18 59 21	1 291 467 969	33.0000000	.2882765	77	1 38 53 29	1 630 532 233	.3074336	.5582552
1090	1 18 81 00	1 295 029 000	33.0151480	10.2914247	78	1 38 76 84	1 634 691 752	.3220046	.5612445
91	1 19 02 81	1 298 596 571	.0302891	.2945709	79	1 39 00 41	1 638 858 339	.3365694	.5642322
92	1 19 24 64	1 302 170 688	.0454233	.2977153	1180	1 39 24 00	1 643 032 000	34.3511281	10.5672181
93	1 19 46 49	1 305 751 357	.0605505	.3008577	81	1 39 47 61	1 647 212 741	.3656805	.5702024
94	1 19 68 36	1 309 338 584	.0756708	.3039982	82	1 39 71 24	1 651 400 568	.3802268	.5731849
1095	1 19 90 25	1 312 932 375	33.0907842	10.3071368	83	1 39 94 89	1 655 595 487	.3947670	.5761658
96	1 20 12 16	1 316 532 736	.1058907	.3102735	84	1 40 18 56	1 659 797 504	.4093011	.5791449
97	1 20 34 09	1 320 139 673	.1209903	.3134083	1185	1 40 42 25	1 664 006 625	34.4238289	10.5821225
98	1 20 56 04	1 323 753 192	.1360830	.3165411	86	1 40 65 96	1 668 222 856	.4383507	.5850983
99	1 20 78 01	1 327 373 299	.1511689	.3196721	87	1 40 89 69	1 672 446 203	.4528663	.5880725
1100	1 21 00 00	1 331 000 000	33.1662479	10.3228012	88	1 41 13 44	1 676 676 672	.4673759	.5910450
1	1 21 22 01	1 334 633 301	.1813200	.3259284	89	1 41 37 21	1 680 914 269	.4818793	.5940158
2	1 21 44 04	1 338 273 208	.1963853	.3290537	1190	1 41 61 00	1 685 159 000	34.4963766	10.5969850
3	1 21 66 09	1 341 919 727	.2114438	.3321770	91	1 41 84 81	1 689 410 871	.5108678	.5999525
4	1 21 88 16	1 345 572 864	.2264955	.3352985	92	1 42 08 64	1 693 669 888	.5253530	.6029184
1105	1 22 10 25	1 349 232 625	33.2415403	10.3384181	93	1 42 32 49	1 697 936 057	.5398321	.6058826
6	1 22 32 36	1 352 899 016	.2565783	.3415358	94	1 42 56 36	1 702 209 384	.5543051	.6088451
7	1 22 54 49	1 356 572 043	.2716095	.3446517	1195	1 42 80 25	1 706 489 875	34.5687720	10.6118060
8	1 22 76 64	1 360 251 712	.2866339	.3477657	96	1 43 04 16	1 710 777 536	.5832329	.6147652
9	1 22 98 81	1 363 938 029	.3016516	.3508778	97	1 43 28 09	1 715 072 373	.5976879	.6177228
1110	1 23 21 00	1 367 631 000	33.3166625	10.3539880	98	1 43 52 04	1 719 374 392	.6121366	.6206788
11	1 23 43 21	1 371 330 631	.3316666	.3570964	99	1 43 76 01	1 723 683 599	.6265794	.6236331
12	1 23 65 44	1 375 036 928	.3466640	.3602029	1200	1 44 00 00	1 728 000 000	34.6410162	10.6265857
13	1 23 87 69	1 378 749 897	.3616546	.3633076	1	1 44 24 01	1 732 323 601	.6554469	.6295367
14	1 24 09 96	1 382 469 544	.3766385	.3664103	2	1 44 48 04	1 736 654 408	.6698716	.6324860
1115	1 24 32 25	1 386 195 875	33.3916157	10.3695113	3	1 44 72 09	1 740 992 427	.6842904	.6354338
16	1 24 54 56	1 389 928 896	.4065862	.3726103	4	1 44 96 16	1 745 337 664	.6987031	.6383799
17	1 24 76 89	1 393 668 613	.4215499	.3757076	1205	1 45 20 25	1 749 690 125	34.7131099	10.6413244
18	1 24 99 24	1 397 415 032	.4365070	.3788030	6	1 45 44 36	1 754 049 816	.7275107	.6442672
19	1 25 21 61	1 401 168 159	.4514573	.3818965	7	1 45 68 49	1 758 416 743	.7419055	.6472085
1120	1 25 44 00	1 404 928 000	33.4664011	10.3849882	8	1 45 92 64	1 762 790 912	.7562944	.6501480
21	1 25 66 41	1 408 694 561	.4813381	.3880781	9	1 46 16 81	1 767 172 329	.7706773	.6530860
22	1 25 88 84	1 412 467 848	.4962684	.3911661	1210	1 46 41 00	1 771 561 000	34.7850543	10.6560223
23	1 26 11 29	1 416 247 867	.5111922	.3942523	11	1 46 65 21	1 775 956 931	.7994253	.6589570
24	1 26 33 76	1 420 034 624	.5261092	.3973366	12	1 46 89 44	1 780 360 128	.8137904	.6618902
1125	1 26 56 25	1 423 828 125	33.5410196	10.4004192	13	1 47 13 69	1 784 770 597	.8281495	.6648217
26	1 26 78 76	1 427 628 376	.5559234	.4034999	14	1 47 37 96	1 789 188 344	.8425028	.6677516
27	1 27 01 29	1 431 435 383	.5708206	.4065787	1215	1 47 62 25	1 793 613 375	34.8568501	10.6706799
28	1 27 23 84	1 435 249 152	.5857112	.4096557	16	1 47 86 56	1 798 045 696	.8711915	.6736066
29	1 27 46 41	1 439 069 689	.6005952	.4127310	17	1 48 10 89	1 802 485 313	.8855271	.6765317
1130	1 27 69 00	1 442 897 000	33.6154726	10.4158044	18	1 48 35 24	1 806 932 232	.8998567	.6794552
31	1 27 91 61	1 446 731 091	.6303434	.4188760	19	1 48 59 61	1 811 386 459	.9141805	.6823771
32	1 28 14 24	1 450 571 968	.6452077	.4219458	1220	1 48 84 00	1 815 848 000	34.9284984	10.6852973
33	1 28 36 89	1 454 419 637	.6600653	.4250138	21	1 49 08 41	1 820 316 861	.9428104	.6882160
34	1 28 59 56	1 458 274 104	.6749165	.4280800	22	1 49 32 84	1 824 793 048	.9571166	.6911331
1135	1 28 82 25	1 462 135 375	33.6897610	10.4311443	23	1 49 57 29	1 829 276 567	.9714169	.6940486
36	1 29 04 96	1 466 003 456	.7045991	.4342069	24	1 49 81 76	1 833 767 424	.9857114	.6969625
37	1 29 27 69	1 469 878 353	.7194306	.4372677	1225	1 50 06 25	1 838 265 625	35.0000000	10.6998748
38	1 29 50 44	1 473 760 072	.7342556	.4403267	26	1 50 30 76	1 842 771 176	.0142828	.7027855
39	1 29 73 21	1 477 648 619	.7490741	.4433839	27	1 50 55 29	1 847 284 083	.0285598	.7056947
1140	1 29 96 00	1 481 544 000	33.7638860	10.4464393	28	1 50 79 84	1 851 804 352	.0428309	.7086023

No.	Square	Cube	Sq. Rt.	Cu. Rt.	No.	Square	Cube	Sq. Rt.	Cu. Rt.
28	1 50 79 84	1 851 804 352	.0428309	.7086023	16	1 73 18 56	2 279 122 496	.2767143	.9585215
29	1 51 04 41	1 856 331 989	.0570963	.7115083	17	1 73 44 89	2 284 322 013	.2904946	.9612965
1230	1 51 29 00	1 860 867 000	35.0713558	10.7144127	18	1 73 71 24	2 289 529 432	.3042697	.9640701
31	1 51 53 61	1 865 409 391	.0856096	.7173155	19	1 73 97 61	2 294 744 759	.3180396	.9668423
32	1 51 78 24	1 869 959 168	.0998575	.7202168	1320	1 74 24 00	2 299 968 000	36.3318042	10.9696131
33	1 52 02 89	1 874 516 337	.1140997	.7231165	21	1 74 50 41	2 305 199 161	.3455637	.9723825
34	1 52 27 56	1 879 080 904	.1283361	.7260146	22	1 74 76 84	2 310 438 248	.3593179	.9751505
1235	1 52 52 25	1 883 652 875	35.1425668	10.7289112	23	1 75 03 29	2 315 685 267	.3730670	.9779171
36	1 52 76 96	1 888 232 256	.1567917	.7318062	24	1 75 29 76	2 320 940 224	.3868108	.9806823
37	1 53 01 69	1 892 819 053	.1710108	.7346997	1325	1 75 56 25	2 326 203 125	36.4005494	10.9834462
38	1 53 26 44	1 897 413 272	.1852242	.7375916	26	1 75 82 76	2 331 473 976	.4142829	.9862086
39	1 53 51 21	1 902 014 919	.1994318	.7404819	27	1 76 09 29	2 336 752 783	.4280112	.9889696
1240	1 53 76 00	1 906 624 000	35.2136337	10.7433707	28	1 76 35 84	2 342 039 552	.4417343	.9917293
41	1 54 00 81	1 911 240 521	.2278299	.7462579	29	1 76 62 41	2 347 334 289	.4554523	.9944876
42	1 54 25 64	1 915 864 488	.2420204	.7491436	1330	1 76 89 00	2 352 637 000	36.4691650	10.9972445
43	1 54 50 49	1 920 495 907	.2562051	.7520277	31	1 77 15 61	2 357 947 691	.4828727	11.0000000
44	1 54 75 36	1 925 134 784	.2703842	.7549103	32	1 77 42 24	2 363 266 368	.4965752	.0027541
1245	1 55 00 25	1 929 781 125	35.2845575	10.7577913	33	1 77 68 89	2 368 593 037	.5102725	.0055069
46	1 55 25 16	1 934 434 936	.2987252	.7606708	34	1 77 95 56	2 373 927 704	.5239647	.0082583
47	1 55 50 09	1 939 096 223	.3128872	.7635488	1335	1 78 22 25	2 379 270 375	36.5376518	11.0110082
48	1 55 75 04	1 943 764 992	.3270435	.7664252	36	1 78 48 96	2 384 621 056	.5513338	.0137569
49	1 56 00 01	1 948 441 249	.3411941	.7693001	37	1 78 75 69	2 389 979 753	.5650106	.0165041
1250	1 56 25 00	1 953 125 000	35.3553391	10.7721735	38	1 79 02 44	2 395 346 472	.5786823	.0192500
51	1 56 50 01	1 957 816 251	.3694784	.7750453	39	1 79 29 21	2 400 721 219	.5923489	.0219945
52	1 56 75 04	1 962 515 008	.3836120	.7779156	1340	1 79 56 00	2 406 104 000	36.6060104	11.0247377
53	1 57 00 09	1 967 221 277	.3977400	.7807843	41	1 79 82 81	2 411 494 821	.6196668	.0274795
54	1 57 25 16	1 971 935 064	.4118624	.7836515	42	1 80 09 64	2 416 893 688	.6333181	.0302199
1255	1 57 50 25	1 976 656 375	35.4259792	10.7865173	43	1 80 36 49	2 422 300 607	.6469644	.0329590
56	1 57 75 36	1 981 385 216	.4400903	.7893815	44	1 80 63 36	2 427 715 584	.6606056	.0356967
57	1 58 00 49	1 986 121 593	.4541958	.7922441	1345	1 80 90 25	2 433 138 625	36.6742416	11.0384330
58	1 58 25 64	1 990 865 512	.4682957	.7951053	46	1 81 17 16	2 438 569 736	.6878726	.0411680
59	1 58 50 81	1 995 616 979	.4823900	.7979649	47	1 81 44 09	2 444 008 923	.7014986	.0439017
1260	1 58 76 00	2 000 376 000	35.4964787	10.8008230	48	1 81 71 04	2 449 456 192	.7151195	.0466339
61	1 59 01 21	2 005 142 581	.5105618	.8036797	49	1 81 98 01	2 454 911 549	.7287353	.0493649
62	1 59 26 44	2 009 916 728	.5246393	.8065348	1350	1 82 25 00	2 460 375 000	36.7423461	11.0520945
63	1 59 51 69	2 014 698 447	.5387113	.8093884	51	1 82 52 01	2 465 846 551	.7559519	.0548227
64	1 59 76 96	2 019 487 744	.5527777	.8122404	52	1 82 79 04	2 471 326 208	.7695526	.0575497
1265	1 60 02 25	2 024 284 625	35.5668385	10.8150909	53	1 83 06 09	2 476 813 977	.7831483	.0602752
66	1 60 27 56	2 029 089 096	.5808937	.8179400	54	1 83 33 16	2 482 309 864	.7967390	.0629994
67	1 60 52 89	2 033 901 163	.5949434	.8207876	1355	1 83 60 25	2 487 813 875	36.8103246	11.0657222
68	1 60 78 24	2 038 720 832	.6089876	.8236336	56	1 83 87 36	2 493 326 016	.8239053	.0684437
69	1 61 03 61	2 043 548 109	.6230262	.8264782	57	1 84 14 49	2 498 846 293	.8374809	.0711639
1270	1 61 29 00	2 048 383 000	35.6370593	10.8293213	58	1 84 41 64	2 504 374 712	.8510515	.0738828
71	1 61 54 41	2 053 225 511	.6510869	.8321629	59	1 84 68 81	2 509 911 279	.8646172	.0766003
72	1 61 79 84	2 058 075 648	.6651090	.8350030	1360	1 84 96 00	2 515 456 000	36.8781778	11.0793165
73	1 62 05 29	2 062 933 417	.6791255	.8378416	61	1 85 23 21	2 521 008 881	.8917335	.0820314
74	1 62 30 76	2 067 798 824	.6931366	.8406788	62	1 85 50 44	2 526 569 928	.9052842	.0847449
1275	1 62 56 25	2 072 671 875	35.7071421	10.8435144	63	1 85 77 69	2 532 139 147	.9188299	.0874571
76	1 62 81 76	2 077 552 576	.7211422	.8463485	64	1 86 04 96	2 537 716 544	.9323706	.0901679
77	1 63 07 29	2 082 440 933	.7351367	.8491812	1365	1 86 32 25	2 543 302 125	36.9459064	11.0928775
78	1 63 32 84	2 087 336 952	.7491258	.8520125	66	1 86 59 56	2 548 895 896	.9594372	.0955857
79	1 63 58 41	2 092 240 639	.7631095	.8548422	67	1 86 86 89	2 554 497 863	.9729631	.0982926
1280	1 63 84 00	2 097 152 000	35.7770876	10.8576704	68	1 87 14 24	2 560 108 032	.9864840	.1009982
81	1 64 09 61	2 102 071 041	.7910603	.8604972	69	1 87 41 61	2 565 726 409	37.0000000	.1037025
82	1 64 35 24	2 106 997 768	.8050276	.8633225	1370	1 87 69 00	2 571 353 000	37.0135110	11.1064054
83	1 64 60 89	2 111 932 187	.8189894	.8661464	71	1 87 96 41	2 576 987 811	.0270172	.1091070
84	1 64 86 56	2 116 874 304	.8329457	.8689687	72	1 88 23 84	2 582 630 848	.0405184	.1118073
1285	1 65 12 25	2 121 824 125	35.8468966	10.8717897	73	1 88 51 29	2 588 282 117	.0540146	.1145064
86	1 65 37 96	2 126 781 656	.8608421	.8746091	74	1 88 78 76	2 593 941 624	.0675060	.1172041
87	1 65 63 69	2 131 746 903	.8747822	.8774271	1375	1 89 06 25	2 599 609 375	37.0809924	11.1199004
88	1 65 89 44	2 136 719 872	.8887169	.8802436	76	1 89 33 76	2 605 285 376	.0944740	.1225955
89	1 66 15 21	2 141 700 569	.9026461	.8830587	77	1 89 61 29	2 610 969 633	.1079506	.1252893
1290	1 66 41 00	2 146 689 000	35.9165699	10.8858723	78	1 89 88 84	2 616 662 152	.1214224	.1279817
91	1 66 66 81	2 151 685 171	.9304884	.8886845	79	1 90 16 41	2 622 362 939	.1348893	.1306729
92	1 66 92 64	2 156 689 088	.9444015	.8914952	1380	1 90 44 00	2 628 072 000	37.1483512	11.1333628
93	1 67 18 49	2 161 700 757	.9583092	.8943044	81	1 90 71 61	2 633 789 341	.1618084	.1360514
94	1 67 44 36	2 166 720 184	.9722115	.8971123	82	1 90 99 24	2 639 514 968	.1752606	.1387386
1295	1 67 70 25	2 171 747 375	35.9861084	10.8999186	83	1 91 26 89	2 645 248 887	.1887079	.1414246
96	1 67 96 16	2 176 782 336	36.0000000	.9027235	84	1 91 54 56	2 650 991 104	.2021505	.1441093
97	1 68 22 09	2 181 825 073	.0138862	.9055269	1385	1 91 82 25	2 656 741 625	37.2155881	11.1467926
98	1 68 48 04	2 186 875 592	.0277671	.9083290	86	1 92 09 96	2 662 500 456	.2290209	.1494747
99	1 68 74 01	2 191 933 899	.0416426	.9111296	87	1 92 37 69	2 668 267 603	.2424489	.1521555
1300	1 69 00 00	2 197 000 000	36.0555128	10.9139287	88	1 92 65 44	2 674 043 072	.2558720	.1548350
1	1 69 26 01	2 202 073 901	.0693776	.9167265	89	1 92 93 21	2 679 826 869	.2692903	.1575133
2	1 69 52 04	2 207 155 608	.0832371	.9195228	1390	1 93 21 00	2 685 619 000	37.2827037	11.1601903
3	1 69 78 09	2 212 245 127	.0970913	.9223177	91	1 93 48 81	2 691 419 471	.2961124	.1628659
4	1 70 04 16	2 217 342 464	.1109402	.9251111	92	1 93 76 64	2 697 228 288	.3095162	.1655403
1305	1 70 30 25	2 222 447 625	36.1247837	10.9279031	93	1 94 04 49	2 703 045 457	.3229152	.1682134
6	1 70 56 36	2 227 560 616	.1386220	.9306937	94	1 94 32 36	2 708 870 984	.3363094	.1708852
7	1 70 82 49	2 232 681 443	.1524550	.9334829	1395	1 94 60 25	2 714 704 875	37.3496988	11.1735558
8	1 71 08 64	2 237 810 112	.1662826	.9362706	96	1 94 88 16	2 720 547 136	.3630834	.1762250
9	1 71 34 81	2 242 946 629	.1801050	.9390569	97	1 95 16 09	2 726 397 773	.3764632	.1788930
1310	1 71 61 00	2 248 091 000	36.1939221	10.9418418	98	1 95 44 04	2 732 256 792	.3898382	.1815598
11	1 71 87 21	2 253 243 231	.2077340	.9446253	99	1 95 72 01	2 738 124 199	.4032084	.1842252
12	1 72 13 44	2 258 403 328	.2215406	.9474074	1400	1 96 00 00	2 744 000 000	37.4165738	11.1868894
13	1 72 39 69	2 263 571 297	.2353419	.9501880	1	1 96 28 01	2 749 884 201	.4299345	.1895523
14	1 72 65 96	2 268 747 144	.2491379	.9529673	2	1 96 56 04	2 755 776 808	.4432904	.1922139
1315	1 72 92 25	2 273 930 875	36.2629287	10.9557451	3	1 96 84 09	2 761 677 827	.4566416	.1948743
16	1 73 18 56	2 279 122 496	.2767143	.9585215	4	1 97 12 16	2 767 587 264	.4699880	.1975334

No.	Square	Cube	Sq. Rt.	Cu. Rt.	No.	Square	Cube	Sq. Rt.	Cu. Rt.
4	1 97 12 16	2 767 587 264	.4699880	.1975334	92	2 22 60 64	3 321 287 488	.6264158	.4267556
1405	1 97 40 25	2 773 505 125	37.4833296	11.2001913	93	2 22 90 49	3 327 970 157	.6393582	.4293079
6	1 97 68 36	2 779 431 416	.4966665	.2028479	94	2 23 20 36	3 334 661 784	.6522962	.4318591
7	1 97 96 49	2 785 366 143	.5099987	.2055032	1495	2 23 50 25	3 341 362 375	38.6652299	11.4344092
8	1 98 24 64	2 791 309 312	.5233261	.2081573	96	2 23 80 16	3 348 071 936	.6781592	.4369581
9	1 98 52 81	2 797 260 929	.5366487	.2108101	97	2 24 10 09	3 354 790 473	.6910843	.4395059
1410	1 98 81 00	2 803 221 000	37.5499667	11.2134617	98	2 24 40 04	3 361 517 992	.7040050	.4420525
11	1 99 09 21	2 809 189 531	.5632799	.2161120	99	2 24 70 01	3 368 254 499	.7169214	.4445980
12	1 99 37 44	2 815 166 528	.5765885	.2187611	1500	2 25 00 00	3 375 000 000	38.7298335	11.4471424
13	1 99 65 69	2 821 151 997	.5898922	.2214089	1	2 25 30 01	3 381 754 501	.7427412	.4496857
14	1 99 93 96	2 827 145 944	.6031913	.2240554	2	2 25 60 04	3 388 518 008	.7556447	.4522278
1415	2 00 22 25	2 833 148 375	37.6164857	11.2267007	3	2 25 90 09	3 395 290 527	.7685439	.4547688
16	2 00 50 56	2 839 159 296	.6297754	.2293448	4	2 26 20 16	3 402 072 064	.7814389	.4573086
17	2 00 78 89	2 845 178 713	.6430604	.2319876	1505	2 26 50 25	3 408 862 625	38.7943294	11.4598474
18	2 01 07 24	2 851 206 632	.6563407	.2346292	6	2 26 80 36	3 415 662 216	.8072158	.4623850
19	2 01 35 61	2 857 243 059	.6696164	.2372696	7	2 27 10 49	3 422 470 843	.8200978	.4649215
1420	2 01 64 00	2 863 288 000	37.6828874	11.2399087	8	2 27 40 64	3 429 288 512	.8329757	.4674568
21	2 01 92 41	2 869 341 461	.6961536	.2425465	9	2 27 70 81	3 436 115 229	.8458491	.4699911
22	2 02 20 84	2 875 403 448	.7094153	.2451831	1510	2 28 01 00	3 442 951 000	38.8587184	11.4725242
23	2 02 49 29	2 881 473 967	.7226722	.2478185	11	2 28 31 21	3 449 795 831	.8715834	.4750562
24	2 02 77 76	2 887 553 024	.7359245	.2504527	12	2 28 61 44	3 456 649 728	.8844442	.4775871
1425	2 03 06 25	2 893 640 625	37.7491722	11.2530856	13	2 28 91 69	3 463 512 697	.8973006	.4801169
26	2 03 34 76	2 899 736 776	.7624152	.2557173	14	2 29 21 96	3 470 384 744	.9101529	.4826455
27	2 03 63 29	2 905 841 483	.7756535	.2583478	1515	2 29 52 25	3 477 265 875	38.9230009	11.4851731
28	2 03 91 84	2 911 954 752	.7888873	.2609770	16	2 29 82 56	3 484 156 096	.9358447	.4876995
29	2 04 20 41	2 918 076 589	.8021163	.2636050	17	2 30 12 89	3 491 055 413	.9486841	.4902249
1430	2 04 49 00	2 924 207 000	37.8153408	11.2662318	18	2 30 43 24	3 497 963 832	.9615194	.4927491
31	2 04 77 61	2 930 345 991	.8285606	.2688573	19	2 30 73 61	3 504 881 359	.9743505	.4952722
32	2 05 06 24	2 936 493 568	.8417759	.2714816	1520	2 31 04 00	3 511 808 000	38.9871774	11.4977942
33	2 05 34 89	2 942 649 737	.8549864	.2741047	21	2 31 34 41	3 518 743 761	39.0000000	.5003151
34	2 05 63 56	2 948 814 504	.8681924	.2767266	22	2 31 64 84	3 525 688 648	.0128184	.5028348
1435	2 05 92 25	2 954 987 875	37.8813938	11.2793472	23	2 31 95 29	3 532 642 667	.0256326	.5053535
36	2 06 20 96	2 961 169 856	.8945906	.2819666	24	2 32 25 76	3 539 605 824	.0384426	.5078711
37	2 06 49 69	2 967 360 453	.9077828	.2845849	1525	2 32 56 25	3 546 578 125	39.0512483	11.5103876
38	2 06 78 44	2 973 559 672	.9209704	.2872019	26	2 32 86 76	3 553 559 576	.0640499	.5129030
39	2 07 07 21	2 979 767 519	.9341535	.2898177	27	2 33 17 29	3 560 550 183	.0768473	.5154173
1440	2 07 36 00	2 985 984 000	37.9473319	11.2924323	28	2 33 47 84	3 567 549 952	.0896406	.5179305
41	2 07 64 81	2 992 209 121	.9605058	.2950457	29	2 33 78 41	3 574 558 889	.1024296	.5204425
42	2 07 93 64	2 998 442 888	.9736751	.2976579	1530	2 34 09 00	3 581 577 000	39.1152144	11.5229535
43	2 08 22 49	3 004 685 307	.9868398	.3002688	31	2 34 39 61	3 588 604 291	.1279951	.5254634
44	2 08 51 36	3 010 936 384	38.0000000	.3028786	32	2 34 70 24	3 595 640 768	.1407716	.5279722
1445	2 08 80 25	3 017 196 125	38.0131556	11.3054871	33	2 35 00 89	3 602 686 437	.1535439	.5304799
46	2 09 09 16	3 023 464 536	.0263067	.3080945	34	2 35 31 56	3 609 741 304	.1663120	.5329865
47	2 09 38 09	3 029 741 623	.0394532	.3107006	1535	2 35 62 25	3 616 805 375	39.1790760	11.5354920
48	2 09 67 04	3 036 027 392	.0525952	.3133056	36	2 35 92 96	3 623 878 656	.1918359	.5379965
49	2 09 96 01	3 042 321 849	.0657326	.3159094	37	2 36 23 69	3 630 961 153	.2045915	.5404998
1450	2 10 25 00	3 048 625 000	38.0788655	11.3185119	38	2 36 54 44	3 638 052 872	.2173431	.5430021
51	2 10 54 01	3 054 936 851	.0919939	.3211132	39	2 36 85 21	3 645 153 819	.2300905	.5455033
52	2 10 83 04	3 061 257 408	.1051178	.3237134	1540	2 37 16 00	3 652 264 000	39.2428337	11.5480034
53	2 11 12 09	3 067 586 677	.1182371	.3263124	41	2 37 46 81	3 659 383 421	.2555728	.5505025
54	2 11 41 16	3 073 924 664	.1313519	.3289102	42	2 37 77 64	3 666 512 088	.2683078	.5530004
1455	2 11 70 25	3 080 271 375	38.1444622	11.3315067	43	2 38 08 49	3 673 650 007	.2810387	.5554973
56	2 11 99 36	3 086 626 816	.1575681	.3341022	44	2 38 39 36	3 680 797 184	.2937654	.5579931
57	2 12 28 49	3 092 990 993	.1706693	.3366964	1545	2 38 70 25	3 687 953 625	39.3064880	11.5604878
58	2 12 57 64	3 099 363 912	.1837662	.3392894	46	2 39 01 16	3 695 119 336	.3192065	.5629815
59	2 12 86 81	3 105 745 579	.1968585	.3418813	47	2 39 32 09	3 702 294 323	.3319208	.5654740
1460	2 13 16 00	3 112 136 000	38.2099463	11.3444719	48	2 39 63 04	3 709 478 592	.3446311	.5679655
61	2 13 45 21	3 118 535 181	.2230297	.3470614	49	2 39 94 01	3 716 672 149	.3573373	.5704559
62	2 13 74 44	3 124 943 128	.2361085	.3496497	1550	2 40 25 00	3 723 875 000	39.3700394	11.5729453
63	2 14 03 69	3 131 359 847	.2491829	.3522368	51	2 40 56 01	3 731 087 151	.3827373	.5754336
64	2 14 32 96	3 137 785 344	.2622529	.3548227	52	2 40 87 04	3 738 308 608	.3954312	.5779208
1465	2 14 62 25	3 144 219 625	38.2753184	11.3574075	53	2 41 18 09	3 745 539 377	.4081210	.5804069
66	2 14 91 56	3 150 662 696	.2883794	.3599911	54	2 41 49 16	3 752 779 464	.4208067	.5828919
67	2 15 20 89	3 157 114 563	.3014360	.3625735	1555	2 41 80 25	3 760 028 875	39.4334883	11.5853759
68	2 15 50 24	3 163 575 232	.3144881	.3651547	56	2 42 11 36	3 767 287 616	.4461658	.5878588
69	2 15 79 61	3 170 044 709	.3275358	.3677347	57	2 42 42 49	3 774 555 693	.4588393	.5903407
1470	2 16 09 00	3 176 523 000	38.3405790	11.3703136	58	2 42 73 64	3 781 833 112	.4715087	.5928215
71	2 16 38 41	3 183 010 111	.3536178	.3728914	59	2 43 04 81	3 789 119 879	.4841740	.5953013
72	2 16 67 84	3 189 506 048	.3666522	.3754679	1560	2 43 36 00	3 796 416 000	39.4968353	11.5977799
73	2 16 97 29	3 196 010 817	.3796821	.3780433					
74	2 17 26 76	3 202 524 424	.3927076	.3806175					
1475	2 17 56 25	3 209 046 875	38.4057287	11.3831906					
76	2 17 85 76	3 215 578 176	.4187454	.3857625					
77	2 18 15 29	3 222 118 333	.4317577	.3883332					
78	2 18 44 84	3 228 667 352	.4447656	.3909028					
79	2 18 74 41	3 235 225 239	.4577691	.3934712					
1480	2 19 04 00	3 241 792 000	38.4707681	11.3960384					
81	2 19 33 61	3 248 367 641	.4837627	.3986045					
82	2 19 63 24	3 254 952 168	.4967530	.4011695					
83	2 19 92 89	3 261 545 587	.5097390	.4037332					
84	2 20 22 56	3 268 147 904	.5227206	.4062959					
1485	2 20 52 25	3 274 759 125	38.5356977	11.4088574					
86	2 20 81 96	3 281 379 256	.5486705	.4114177					
87	2 21 11 69	3 288 008 303	.5616389	.4139769					
88	2 21 41 44	3 294 646 272	.5746030	.4165349					
89	2 21 71 21	3 301 293 169	.5875627	.4190918					
1490	2 22 01 00	3 307 949 000	38.6005181	11.4216476					
91	2 22 30 81	3 314 613 771	.6134691	.4242022					
92	2 22 60 64	3 321 287 488	.6264158	.4267556					

Answers to Math Problems

CHAPTER 2

Page 14
69; 116; 540; 1,128; 1,233; 5,120; 8,012; 182; 12,007; 25,671

Page 15
Larger than 200 but smaller than 220.

Page 25
225; 373; 2,522; 2,840; 3,758; 24,747; $17.79; $11.49

CHAPTER 3

Page 36
288; 644; 126; 224; 2,548; 3,564; 132,678; 189,288

Page 38
4800; 19,700; 1,440; 42,800; 7,956; 3,198,400,000; 1,775,000

CHAPTER 4

Page 43
8,370; 9,118; 8,832; 9,405; 8,160; 7,656; 10,608; 12,208; 12,305

Page 47
2,205; 2,112; 1,880; 1,554; 2,970; 2,964; 3,480; 3,233

Page 49
728; 504; 616; 506; 725; 792; 1,184; 999; 1,326

Page 51
5,254; 4,526; 5,400; 6,162; 4,964; 4,690; 6,396; 6,320; 4,736

Page 53
47,515; 86,724; 150,508; 241,560; 266,240; 482,318; 993,010

CHAPTER 5

Page 63
392; 10,296; 55,566; 129,558; 150,080; 2,177,355

CHAPTER 6

Page 69
8,836; 7,921; 12,769; 7,056; 11,881; 16,129

Page 71
986,049; 996,004; 1,008,016; 996,004; 1,012,036; 952,576; 1,024,144

Page 75
Rows, horizontally: 225; 7,225; 11,025; 119,025; 1,525,225; 308,025; 961; 5,041; 12,321; 2,601; 3,721; 8,281; 33,856; 196; 4,096; 7,056; 2,916; 12,996; 3,136; 11,236; 9,216; 5,776; 256; 4,356; 361; 9,801; 4,761; 11,881; 3,481; 1,521
Word problem: $24^2 - 24 = 552$ boxes of Chinese soup mix. This assumes all the boxes in every row above the bottom tumble to the floor.

Page 79
1,672; 216; 572; 3,965; 304; 6,804; 728; 5,624; 210; 2,652

CHAPTER 7

Page 83

4,820; 367,200; 19,960; 82,800; 8,640; 1,350,000

$421^{10\times9}$; $9{,}876^{10\times9}$; $18^{10\times12}$; $47^{10\times12}$; $13^{10\times15}$; $52^{10\times100}$

Page 87

198; 492; 837; 4,026; 4,182; 2,604

Pages 89, 90

2; 4; 13; 27; 54

10; 28; 117; 297; 702

60; 540; 2,870; 1,980; 1,530

Page 93

210; 1,125; 1,710; 1,035; 2,610; 6,390; 1,045; 3,520; 6,655

Page 96

684; 2,268; 486; 3,069; 6,930; 6,831; 32,076

CHAPTER 8

Page 98

9; 12; 49; 40; 50; 102

Page 100

6; 6; 16; 18; 18; 28

Page 102

329; 115; 206; 172; 139; 79

Page 103

8; −8; 25; −25; 114; −114

Page 107

88; 25; 144; 66; 344; 3,245

Page 108

35; 39; 51; 125; 250

CHAPTER 9

Page 115

How you choose to work these problems depends on your taste and your skill. Answers: 20.6; 4760.6; 7.16; 7207.91; 308.069

Page 118

12%; 25%; 36%; 5%; 100%
2487.6; 4,868.2732; 92,367.4163; .1248; 4.86; .98

Page 121

2.8; 57.3; 5.2; 3

CHAPTER 10

Page 124

61.8; 36.08; 6.832; 14.4; 1.724; 38.944

Pages 132, 133

Divisible by 3: 126; 1,704; 168; 171
Divisible by 4: 96; 236; 944; 3,780
Divisible by 5: 15,670; 42,345; 12,000
Divisible by 6: 270; 3,954; 3,996; 25,452
Divisible by 8: 576; 31,536
Divisible by 9: 369; 693; 963; 8,847
Divisible by 11: 9,856; 7,194
Divisible by 12: 9,468; 5,472; 11,556
Divisible by 15: 11,115; 2,310; 945; 147,795

$\begin{array}{r} 951 \\ 25\overline{)23{,}775} \end{array}$	$\begin{array}{r} 63 \\ 8\overline{)504} \end{array}$	$\begin{array}{r} 5{,}874 \\ 22\overline{)29{,}888} \end{array}$
$\begin{array}{r} 4{,}356 \\ 12\overline{)52{,}272} \end{array}$	$\begin{array}{r} 1{,}457 \\ 6\overline{)8{,}642} \end{array}$	$\begin{array}{r} 5{,}874 \\ 11\overline{)64{,}614} \end{array}$
$\begin{array}{r} 987{,}436 \\ 3\overline{)2{,}962{,}368} \end{array}$	$\begin{array}{r} 109 \\ 5\overline{)545} \end{array}$	$\begin{array}{r} 963 \\ 33\overline{)31{,}779} \end{array}$
$\begin{array}{r} 85 \\ 15\overline{)1{,}275} \end{array}$	$\begin{array}{r} 753 \\ 30\overline{)22{,}590} \end{array}$	$\begin{array}{r} 624 \\ 24\overline{)14{,}976} \end{array}$
$\begin{array}{r} 587 \\ 12\overline{)7{,}044} \end{array}$	$\begin{array}{r} 484{,}287 \\ 8\overline{)3{,}874{,}296} \end{array}$	$\begin{array}{r} 96{,}358 \\ 20\overline{)1{,}927{,}160} \end{array}$
$\begin{array}{r} 754 \\ 9\overline{)6{,}786} \end{array}$		

If a number ends in 0 and is evenly divisible by 9, it is also evenly divisible by 90. For example:

$$9\overline{)450} = 50 \qquad 90\overline{)450} = 5$$

CHAPTER 12

Page 147
89; 92; 44; 18; 66; 91; 97; 17

Page 152
Cube roots: 28; 57; 98
Fourth roots: 4; 21; 87
Sixth roots: 6; 11; 32
Ninth root: 2

CHAPTER 13

Page 155
11/4; 19/5; 123/12; 78/8; 525/32

Page 157
16/40 and 20/40; 3/4 and 2/4; 21/30 and 4/30; 60/90 and 72/90, or 30/45 and 36/45; 40/60, 48/60, and 45/60; 396/660, 540/660, and 550/660

Page 159
3/8; 19/24; 1 13/18; 1 23/60; 1 31/40; 61/72; 1 7/80

Page 161
1 11/36; 29/30; 1 67/120; 7 19/40; 11 33/54; 1 88/195; 5/6; 17/72; 1/72

Page 162
2/5; 1/20; 1/8; 11/40; 1/2

Page 164
3 15/28; 10 7/10; 3 3/4; 3 86/165; 1 61/80

Page 168
1/3; 5/14; 10/21; 3 97/270; 1/10; 2/13

Page 170
15; 1/2; 4/5; 4/35; 4 11/16; 1/9

Page 172

4/10 or .4; 15/100 or .15; 27/1,000 or .027; 80/100 or .80; 164/10,000 or .0164; 47/1,000,000 or .000047

Any decimal followed by the appropriate number of zeros will do. An example: .01; .010; .0100; .01000

Page 174

1,134.605; 212.255; 567.4909; 50.896; 55.3

Page 176

4056.1; 37.0736; 4665.3642; 34.713; 3.0955; 195; 515.46; .0776

Page 178

.9; .75; .75; .185; .533; .625

3/4; 5/10; 7/8

CHAPTER 14

Page 181

80%; 88.8%; 25%; 87.5%; 62.5%; 45%; 211.6%; 50.5%; 211.1%

Page 182

48%; 22%; 50%; 87%; 225%

Page 183

9/50; 1/3; 1/10; 9/10; 41/50

Pages 186, 187

7.3%; 64; 16.6%; 45; 33.33%; 70.38

Page 189

5 percent of $10.00 = 50 cents. Add this to the principal to get the total owed after one month: $10.50.

At 5 percent interest compounded weekly over slightly more than four weeks, $1.00 would earn the lender about $1.22. Multiply by $10.00 to get the total you would owe to your kid brother: $12.22.

CHAPTER 15

Pages 195, 196

.0017; 39; 34,890; 12,600; 67; 2,389,476.234; 80; 100

Page 197
15,100; 919.15; 9,543,800,000; 1,944.3 square feet; 2,498; 31.83

Page 199
The answer depends on your choice of grocery items, the current prices, and the amount you're inclined to eat. However, if you have much money left over for a restaurant meal, you probably ate a lot of potatoes during the week.

CHAPTER 16

Page 209
43; 99; 1,821; 25,887; 16,486; 58,025

Page 211
92,729 (correct answer is 29,729)
168,642 (correct answer is 167,642)
690,613 (correct answer is 608,623)
21,797,246 (correct answer is 450,831,146)

Page 214
The only correct answer is $27 \times 23 = 621$. These are the correct answers: 2,484; 621; 36.08; 221,778; 17,659,219; 40,885,120; 635,171,674,722

Page 217
Correct answers: 52; 62, r. 17; 423.52; 526, r. 26; 34; 83.68

CHAPTER 17

Page 224
10 million
1,000,000,000; 1,000,000; 10,000
8,764,522,330; 87,645,223.30

Page 226
Your brother is running a temperature of 99.32. Since a normal temperature is 98.6, he may have a slight fever, but he's not very sick.

CHAPTER 18

Page 231

Sunday; Wednesday; Tuesday; Saturday; Wednesday

CHAPTER 19

Page 238

You round off your tips to the nearest 5 or 10 cents, so that you're not handing out pennies. All told, you spend $5.65 in tips.

You tip the following:

Cabbie:	$2.25
Bellboy:	$1.50
	$1.90
	$5.65

You could have saved your client some change by doing several things differently:

1. Take the shuttle bus instead of a taxi from the airport to the hotel. Shuttles are cheaper than taxis, and if the driver doesn't carry your luggage, you don't have to tip him.
2. Carry your own bags to your room. This kind of behavior, however, would look tacky in a swell hotel like the Brown Palace. If you stayed in a motel, you would be expected to carry your bags, and you would probably save on the room rate, too.
3. Go down to the coffee shop for a snack, instead of having it expensively delivered to your room.
4. Bring your own cheese and crackers in your luggage and walk down the hall to the soft drink machine for a soft drink.

Page 242

The waiter owed you $19.46.

The total bill is $36.86, out of $50:

36.86 + 1 cent
.87 + 1 cent
.88 + 1 cent
.89 + 1 cent
.90 + 10 cents
37.00 + $1.00
38.00 + $1.00
39.00 + $1.00
40.00 + $10.00
$50.00

Your customer receives $13.14 in change. He gives you a tip of $5.53, or, in round numbers, $5.50.

CHAPTER 20

Page 253
2 hours, 45 minutes
Algiers, Algeria (time depends on your time zone)
1740 hours
An analog clock is circular and so is the Equator of the earth; both the Equator and an analog clock contain 360 degrees; like longitude, time is measured on an analog clock in minutes and seconds; the movement of the hour hand corresponds to the movement of the sun overhead as it passes from meridian to meridian.

6:00 P.M., if by "dinner" we mean the evening meal.

CHAPTER 22

Page 285
$x + y$
$3x$
$348 - 27$
$22 > 11$
$22 = 2x$
$\frac{(93 - 47)}{12}$
$(y + 3)(y - 5)$

Page 287
$14abc$; $14(abc)$; $14a(bc)$; $(14ab)c$; $cba14$; $bca14$; $abc14$; $a(14b)c$; and so forth
$ab \div 75 = 1$
$14x + 14c$
$xy \div 3x - y = 12$
You will get -18.

Page 289
$28x^{10}$; $4c^4$; $104a^7 + 4y$

Page 291
$p = \$1.15 \div 3$
$p = \$1.15 \div 15$

Page 298
35°; −5°
Joe has $30 too little to pay his debt.
−$80
$720
Yes

BOOKS BY SCOTT FLANSBURG

MATH MAGIC
How to Master Everyday Math Problems
ISBN 0-06-072635-0 (paperback)

Scott Flansburg's heartfelt belief is that there are no "mathematical illiterates," just people who have not learned how to make math work for them. In *Math Magic* he demonstrates how to put math-phobias to rest and deal with essential everyday mathematical calculations with confidence. Flansburg makes math what you may never have imagined it to be: easy and fun.

MATH MAGIC FOR YOUR KIDS
Hundreds of Games and Exercises from the Human Calculator to Make Math Fun and Easy
ISBN 0-06-097731-0 (paperback)

With entertaining games and tricks, Flansburg helps kids develop a positive attitude about numbers, the necessary foundation on which they will build math skills for the rest of their education. Children will discover hours of independent amusement, while parents will find activities they can do with their children to supplement their schoolwork and improve grades.

Don't miss the next book by your favorite author.
Sign up for AuthorTracker by visiting *www.AuthorTracker.com*.

Available wherever books are sold, or call 1-800-331-3761 to order.